Systems Design 2.0

--The Structure-Behavior Coalescence Approach--

William S. Chao

Structure-Behavior Coalescence

Systems Architecture = Systems Structure + Systems Behavior

CONTENTS

PREFACE

Systems design (hardware/software codesign) is, in the system development process, the design and implementation phase. That is, systems design means to get a solution to furnish customers' requirements on the system. When working on the systems design, we mainly consider how to manufacture the system, but not to specify what this system is.

A system has been designed, by systems design 1.0, hopefully to be an integrated whole, embodied in its assembled components, their interactions with each other and the environment. Since systems structure and systems behavior are the two most prominent views of a system, integrating the systems structure and systems behavior apparently is the best way to achieve a truly integrated whole of a system. Because systems design 1.0 does not design the integration of systems structure and systems behavior, very likely it will never be able to actually form an integrated whole of a system.

Structure-behavior coalescence (SBC) provides an elegant way to integrate the systems structure and systems behavior, and hence achieves a truly integrated whole, of a system. A truly integrated whole sets a path to achieve the desired systems design. SBC facilitates an integrated whole. Therefore, we conclude that systems design 2.0 using the SBC approach, which contains three fundamental diagrams: a) framework diagram, b) component operation diagram and c) interaction flow diagram, is highly adequate in designing a system.

ABOUT THE AUTHOR

Dr. William S. Chao is the CEO & founder of SBC Architecture International®. SBC (Structure-Behavior Coalescence) architecture is a systems architecture which demands the integration of systems structure and systems behavior of a system. SBC architecture applies to hardware architecture, software architecture, enterprise architecture, knowledge architecture and thinking architecture. The core theme of SBC architecture is: "Architecture = Structure + Behavior."

William S. Chao received his bachelor degree (1976) in telecommunication engineering and master degree (1981) in information engineering, both from the National Chiao-Tung University, Taiwan. From 1976 till 1983, he worked as an engineer at Chung-Hwa Telecommunication Company, Taiwan.

William S. Chao received his master degree (1985) in information science and Ph.D. degree (1988) in information science, both from the University of Alabama at Birmingham, USA. From 1988 till 1991, he worked as a computer scientist at GE Research and Development Center, Schenectady, New York, USA.

ABOUT THE AUTHOR

PART I: BASIC CONCEPTS

Chapter 1: Introduction to Systems Design

Systems design (hardware/software codesign) is part of the overall systems development process as reflected in the systems development life cycle. The phases of the systems development life cycle are as follows: a) project planning, b) requirements and specifications, c) design and implementation, d) verification and validation and e) product evolution. The system life cycle applies recursively to life cycles that produce hardware and software portions of the system.

Systems design is, in the systems development process, the design and implementation phase. That is, systems design means to get a solution to furnish customers' requirements on the system. When working on the systems design, we mainly consider how to manufacture the system, but not to specify what this system is.

1-1 Systems Development Life Cycle

The phases of the systems development life cycle (SDLC) [Blan08, Koss11, Wald15] , as shown in Figure 1-1, are: a) project planning, b) requirements and specifications, c) design and implementation, d) verification and validation and e) product evolution. The systems development life cycle applies recursively to life cycles that produce hardware and software portions of the system.

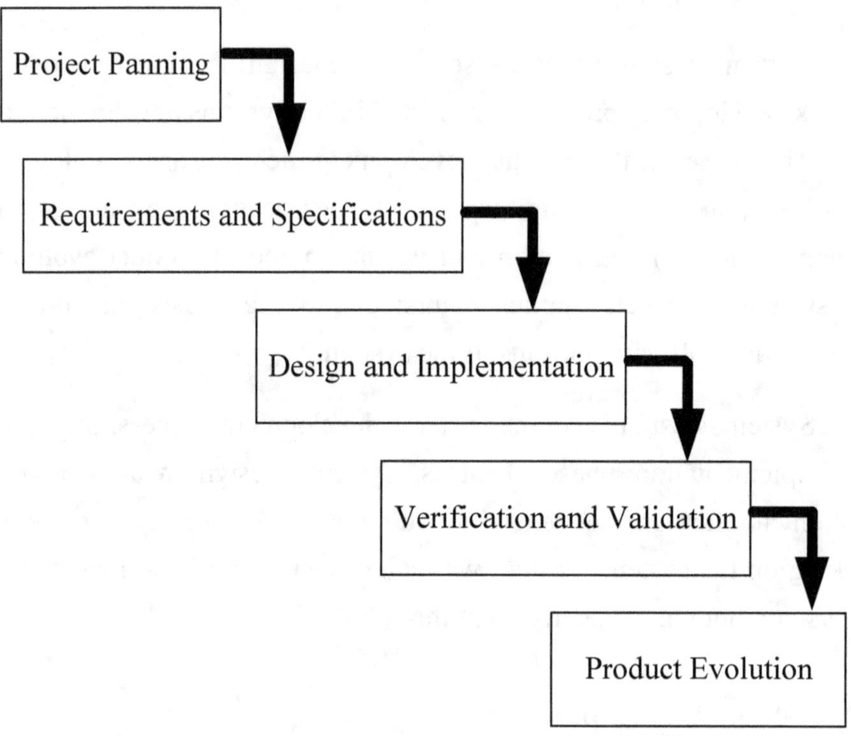

Figure 1-1 Five Phases of the Systems Development Life Cycle

1-1-1 Project Planning

Project planning determines the general goals of the systems development project. These general goals include: project scope determination; selection of the systems process model; selection of the systems engineering development technology; estimating applicable resources; determining systems metric methodology; cost estimation; risk management; project scheduling and tracking; determining the configuration management approach; understanding the level of quality management; choosing systems engineering tools; drawing up contracts; and determining post-project follow up.

1-1-2 Requirements and Specifications

The requirements and specifications phase consists of determining what the customer really requires. Requirements and specifications appertain to the problem space. When working on requirements and specifications, we usually only specify what the system is, but never think about how the system shall be manufactured.

1-1-3 Design and Implementation

The design and implementation phase belongs to the solution space. In other words, design and implementation try to secure a solution to meet or exceed customer requirements. It is opposite to requirements and specifications, design and implementation mainly consider how to manufacture the system, but not to specify what the system is.

1-1-4 Verification and Validation

The fourth step is called the verification and validation, abbreviated as V&V, phase. Verification uses proving technology. Validation uses testing technology. After the system product has been manufactured, we use either verification or validation to determine if or not the system product meets the requirements and specifications initially settled.

1-1-5 Product Evolution

Product evolution is the fifth, also the last, phase of system development life cycle. After verification and validation, we hand over the system product for the customer to use. Uses for several years, several month or even several days later, if has the necessity to carry on the next edition, either perceive that some part of wrong, some parts need the reinforcement, either the customer thinks that some places must change the requirements and specifications, even overhauls greatly, then must carry on the product evolution in accordance.

1-2 Systems Design

Systems design (hardware/software codesign) [Beam90, Card11] may refer to all the activities involved in conceptualizing, framing, implementing, commissioning and ultimately modifying complex systems.

Systems design is, in the systems development life cycle, the design and implementation phase. That is, design and implementation endeavor to get a solution to furnish customers' requirements. When working on the systems design, we mainly consider how to manufacture this system, but not to specify what this system is.

During the design and implementation phase, both analysts and designers need to coordinate closely, exchange the opinion fully, finally achieve the systems design document output, as shown in Figure 1-2.

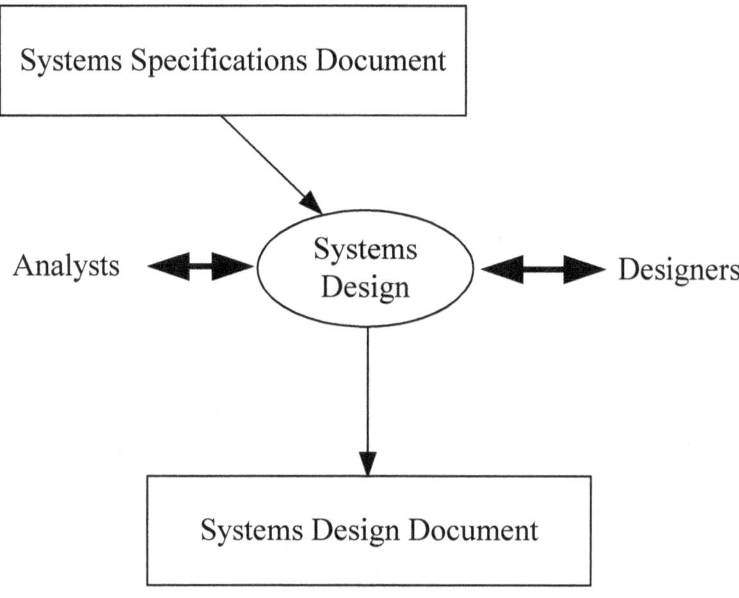

Figure 1-2 Work of Design and Implementation

Analysts maintain notable responsibility in the systems design work. The analyst besides provides the specifications document to the designer, but also needs to exchange the opinion with the designer, to check the rationality of the customer's request.

Designers also uphold grand responsibility in the systems design work. The designer must possess the capability of deriving a sturdy system from the specifications document. In addition, the designer also is accountable to bring together the systems design document.

1-3 Multiple Views of a System

In general, a system is extremely complex that it consists of multiple views such as structure view, behavior view, function view, data view as shown in Figure 1-3 [Denn08, Kend10, Pres09, Somm06].

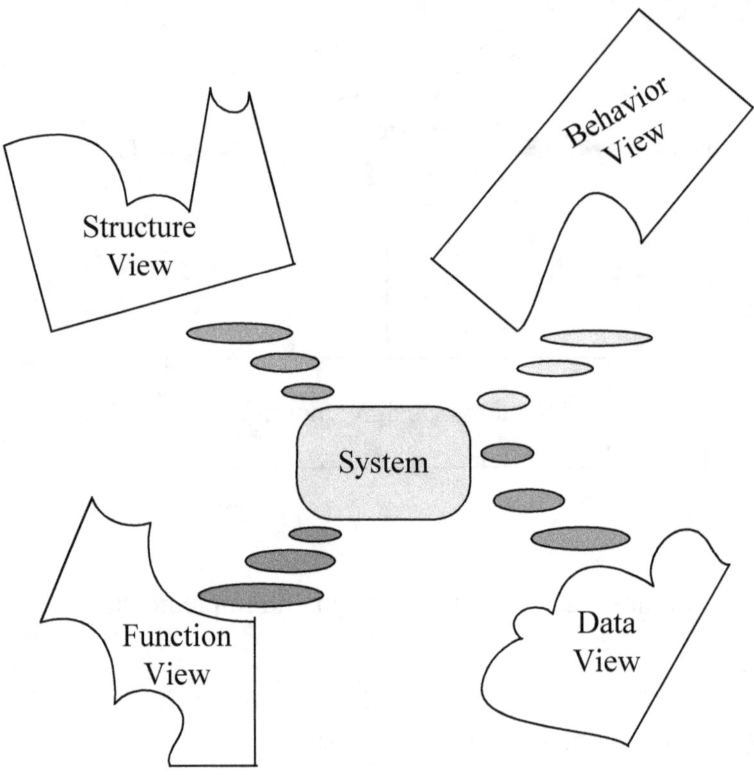

Figure 1-3 Multiple Views of a System

Among the above multiple views, the structure and behavior views are perceived as the two prominent ones. The structure view focuses on the systems structure which is described by components and their composition while the behavior view concentrates on the systems behavior which involves interactions [Chao15a, Chao15b, Chao15c, Chao15d, Chao15e, Hoar85, Miln89, Miln99] among the external environment's actors and components. Function and data views are considered to be other views as shown in Figure 1-4.

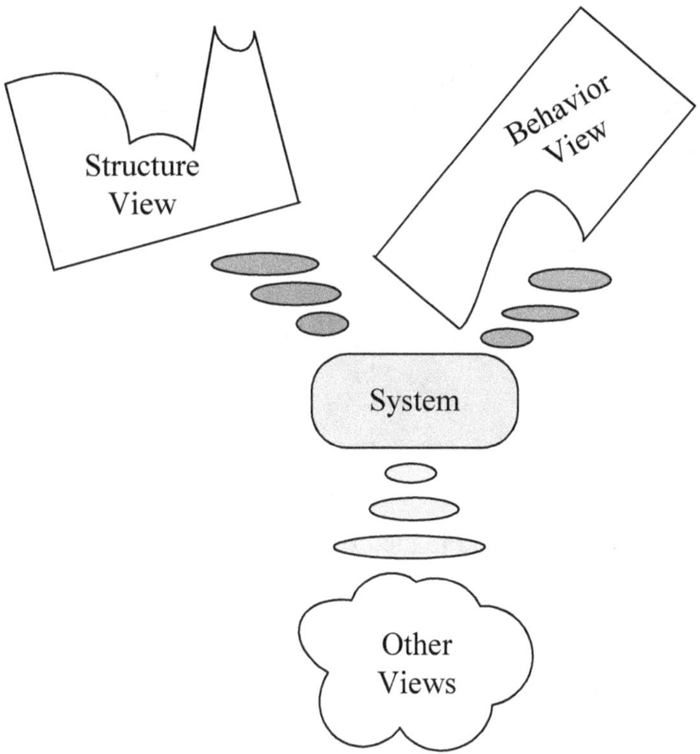

Figure 1-4 Structure, Behavior and Other Views

Either Figure 1-3 or Figure 1-4 represents the multiple views of a system. In some situations Figure 1-3 is used and in other situations Figure 1-4 is used.

Accordingly, a system is designed in Figure 1-5 to be an integrated whole of that system's multiple views, i.e., structure, behavior and other views, embodied in its assembled components, their interactions [Chao15a, Chao15b, Chao15c, Chao15d, Chao15e, Hoar85, Miln89, Miln99] with each other and the environment. Components are sometimes labeled as non-aggregated systems, parts, entities, objects and building blocks [Chao14a, Chao14b, Chao14c].

A system is an integrated whole of that system's multiple views, i.e., structure, behavior, and other views, embodied in its assembled components, their interactions with each other and the environment.

Figure 1-5 Design of a System

Since multiple views are embodied in a system's assembled components which belong to the systems structure, they shall not exist alone. Multiple views must be loaded on the systems structure just like a cargo is loaded on a ship as shown in Figure 1-6. There will be no multiple views if there is no systems structure. Stand-alone multiple views are not meaningful.

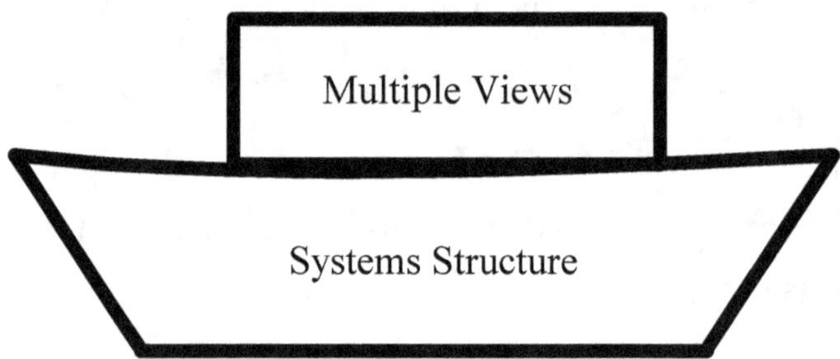

Figure 1-6 Multiple Views Loaded on the Systems Structure

1-4 Multiple Views Non-Integrated Approaches for Systems Design 1.0

When designing a system, the multiple views non-integrated approach, also known as the model multiplicity approach [Dori95, Dori02, Dori16], respectively picks a model for each view as shown in Figure 1-7, the structure view has the structure model; the behavior view has the behavior model; the function view has the function model; the data view has the data model. These multiple models, are heterogeneous and not related to each other, and thus become the primary cause of model multiplicity problems [Dori95, Dori02, Dori16, Pele02, Sode03].

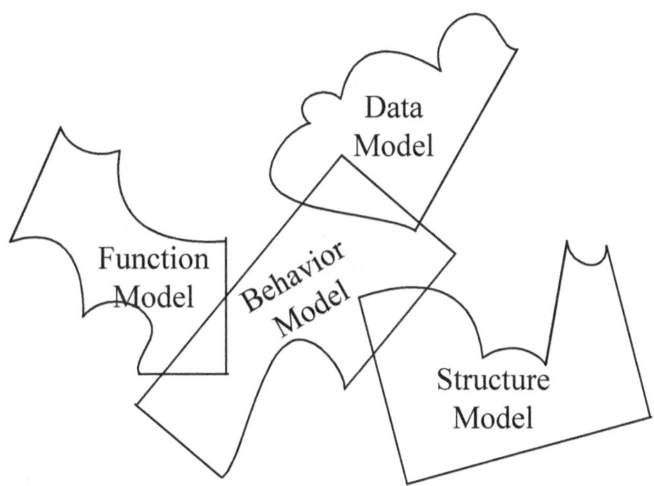

Figure 1-7 Multiple Views Non-Integrated Approach

Multiple views non-integrated approaches for systems design 1.0 fall into four general categories: data-oriented, function-oriented, control-oriented, and object-oriented [Grad13, Hatl00], as shown in Figure 1-8.

Each of these approaches, more or less, fails to design a system as an integrated whole of that system's multiple views.

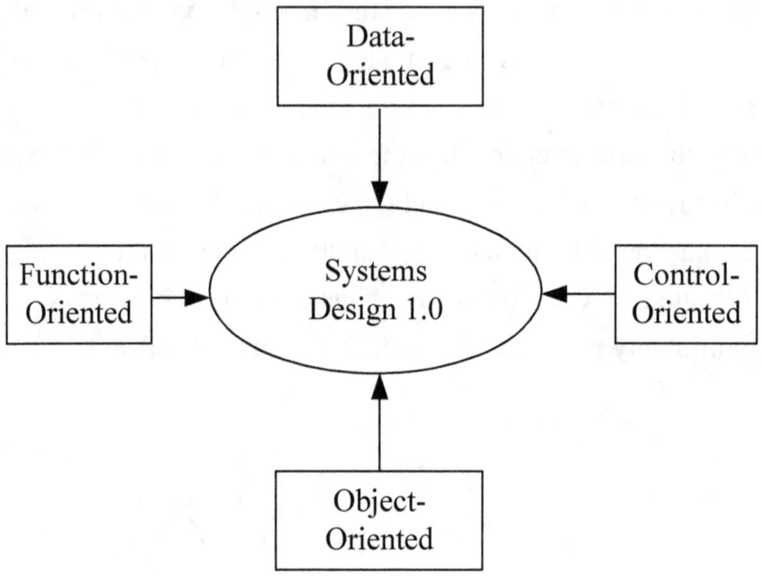

Figure 1-8 Multiple Views Non-Integrated Approaches for Systems Design 1.0

Data-oriented approaches for systems design 1.0 stress the system state as a data structure. Jackson System Development (JSD) [Came89] and Entity Relationship Modeling (ERM) [Chen76] are primarily data-oriented. Data-oriented approaches concentrate only on data and completely neglect to integrate the systems structure and systems behavior. Therefore, data-oriented approaches are multiple views non-integrated and will never become an ideal systems design approach.

Function-oriented approaches for systems design 1.0 take the primary view of the way a system transforms input data into output data. Each transformation from input data into output data demonstrates a function of the system. A system may contain many such kinds of functions which represent the function view of the system. Classical Structured Analysis (SA) [DeMa79] fits into the category of function-oriented approaches, as do Structured Analysis and Design Technique (SADT) [Marc88] and Structured Systems Analysis and Design Method (SSADM) [Ashw90]. Function-oriented approaches concentrate only on the function view and completely neglect to integrate the systems structure and systems behavior. Just like data-oriented approaches, function-oriented approaches are multiple views non-integrated and will never become an ideal systems design approach.

Control-oriented approaches for systems design 1.0 emphasize synchronization, deadlock, exclusion, concurrency and process activation of a system. Petri Net [Reis92] and Flowcharting [Bash86] are primarily control-oriented. Control-oriented approaches concentrate only on the control view and completely neglect an integrated structure and behavior views which grasps the essential properties of a system. Just like data-oriented and function-oriented approaches, control-oriented approaches are multiple views non-integrated and will never become an ideal systems design approach.

Object-oriented approaches for systems design 1.0 design the system as classes of objects and their behaviors. Object-oriented Design (OOD) [Booc07], fitting into the category of object-oriented methods. Object-oriented approaches stress both the structure view and the behavior view, but not an integrated structure and behavior views. Object-oriented approaches do not emphasize to integrate the systems structure and systems behavior. Like data-oriented, function-oriented and control-oriented approaches, object-oriented approaches are multiple views non-integrated and will never become an ideal systems design approach.

1-5 Multiple Views Integrated Approaches for Systems Design 2.0

When designing a system, the multiple views integrated approach, also known as the model singularity approach [Dori95, Dori02, Dori16, Pele02, Sode03], instead of picking many heterogeneous and unrelated models, will use only one single model as shown in Figure 1-9. The structure, behavior, function and data views are all integrated in this one single model which represents an integrated whole of that system's multiple views [Chao14a, Chao14b, Chao14c].

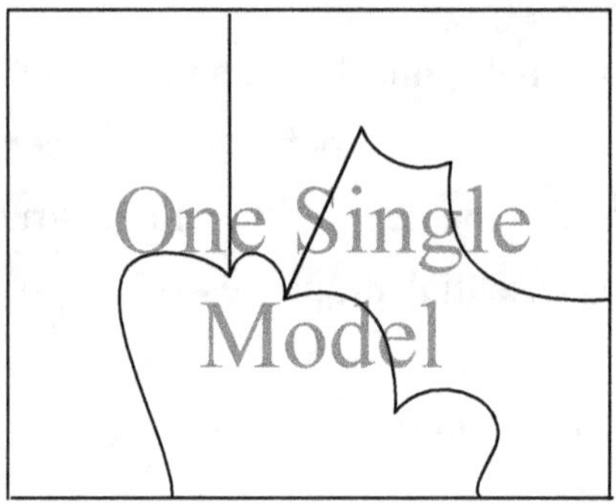

Figure 1-9 Multiple Views Integrated Approach

Multiple views integrated approaches for systems design 2.0 design a system as an integrated whole of that system's multiple views.

Chapter 2: Systems Structure and Systems Behavior

Systems structure and systems behavior are the two most significant views of a system. Systems structure, designed by components, their operations and their composition, refers to the type of connection between the components of a system. Systems behavior, designed by the interactions between and among the components and environment, refers to the interconnectivities a system in conjunction with its environment.

2-1 Structure of Systems

Every system forms a whole. In general, structure of systems is the type of connection between the components of a system. More specifically, we design the structure of a system by 1) components, 2) their operations and 3) their composition.

Components are something relatively indivisible in one system [Hoff10, Shel11]. For example, *New_Account_UI*, *Nearby_Attractions_CityMap_UI*, *Attraction_Details_UI*, *Personalized_Itinerary_UI*, *Checking_In_And_Recommending_UI*, *STCCASIS_Database*, *Tourist_GPS_M (M = AAA0000 to ZZZ9999)* and *Scenic_Spot_RFID_N (N = 000 to 999)* are components of the *Smart Tourism City Cloud Applications and Services IoT System* (STCCASIS) as shown in Figure 2-1.

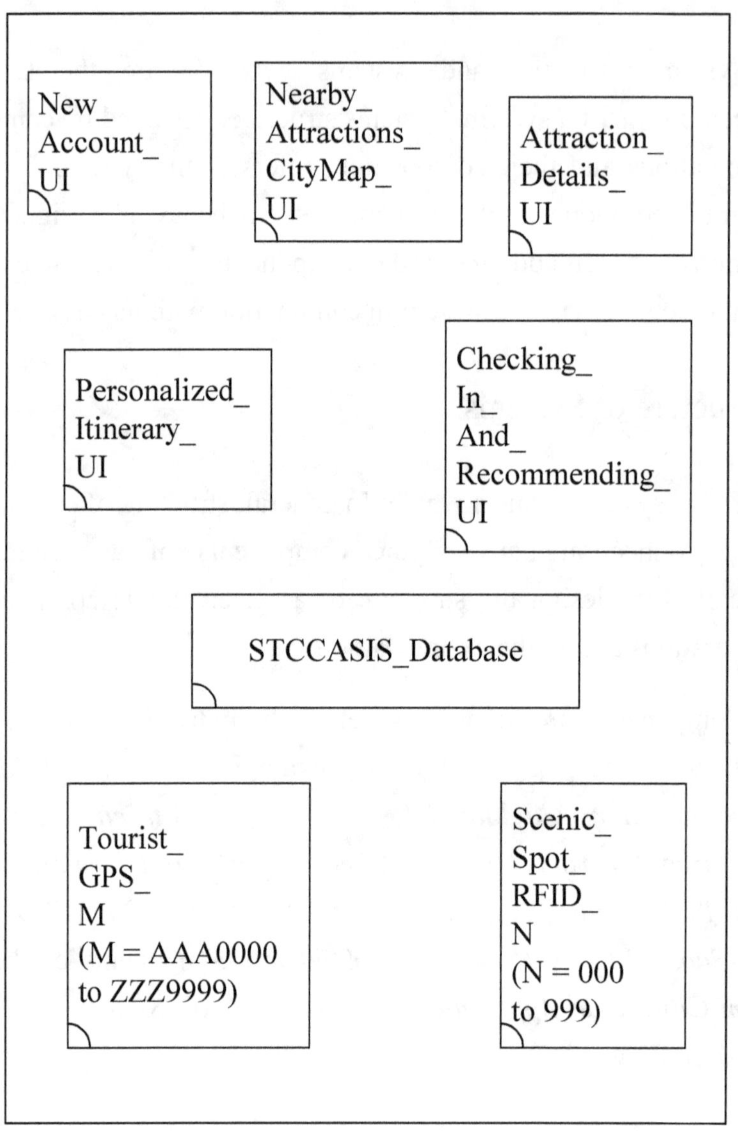

Figure 2-1 Components of the
Smart Tourism City Cloud Applications and Services IoT System

An operation provided by each component represents a procedure or method or function of the component [Chao14a, Chao14b, Chao14c]. Each component in a system must possess at least one operation. Figure 2-2 shows the operations of all components of the *Smart Tourism City Cloud Applications and Services IoT System* (STCCASIS). In the figure, component *New_Account_UI* has one operation: *Input_New_Account*; component *Nearby_Attractions_CityMap_UI* has one operation: *Show_Nearby_Attractions_CityMap*; component *Attraction_Details_UI* has one operation: *Show_Attraction_Details*; component *Personalized_Itinerary_UI* has one operation: *Input_Personalized_Itinerary*; component *Checking_In_And_Recommending_UI* has two operations: *Scenic_Spot_Check_In* and *Scenic_Spot_Recommend*; component *STCCASIS_Database* has five operations: *SQL_Insert_New_Account*, *SQL_Select_Nearby_Attractions*, *SQL_Select_Attraction_Details*, *SQL_Insert_Personalized_Itinerary* and *SQL_Insert_Checking_In_And_Recommending*; component *Tourist_GPS_M (M = AAA0000 to ZZZ9999)* has one operation: *Tourist_GPS_Positioning*; component *Scenic_Spot_RFID_N (N = 000 to 999)* has one operation: *Scenic_Spot_RFID_Positioning*.

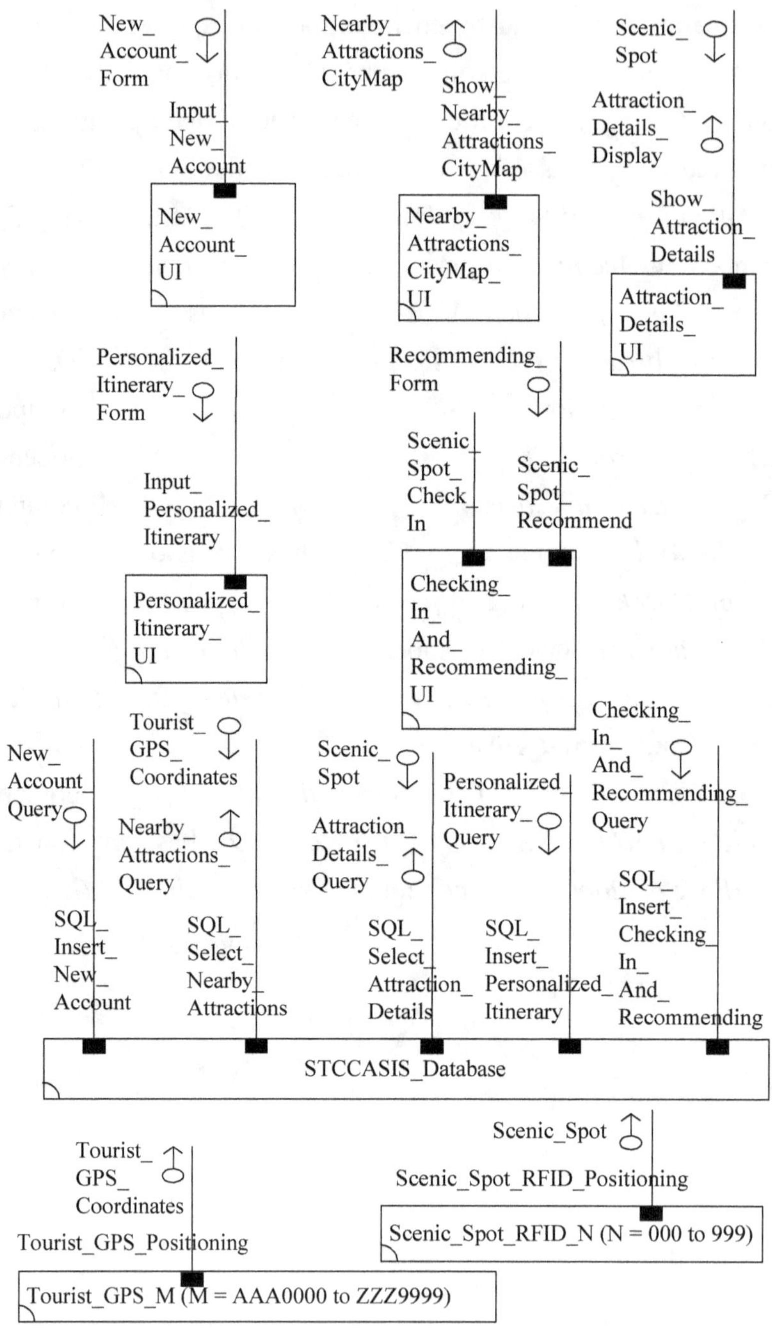

Figure 2-2 Operations of all Components of
the *Smart Tourism City Cloud Applications and Services IoT System*

Composition of components designs the structural composition and decomposition of a system. For example, Figure 2-3 shows that, in the *Smart Tourism City Cloud Applications and Services IoT System* (STCCASIS), *Application_Layer* contains the *New_Account_UI*, *Nearby_Attractions_CityMap_UI*, *Attraction_Details_UI*, *Personalized_Itinerary_UI* and *Checking_In_And_Recommending_UI* components; *Data_Layer* contains the *STCCASIS_Database* component; *Technology_Layer* contains the *Tourist_GPS_M (M = AAA0000 to ZZZ9999)* and *Scenic_Spot_RFID_N (N = 000 to 999)* components.

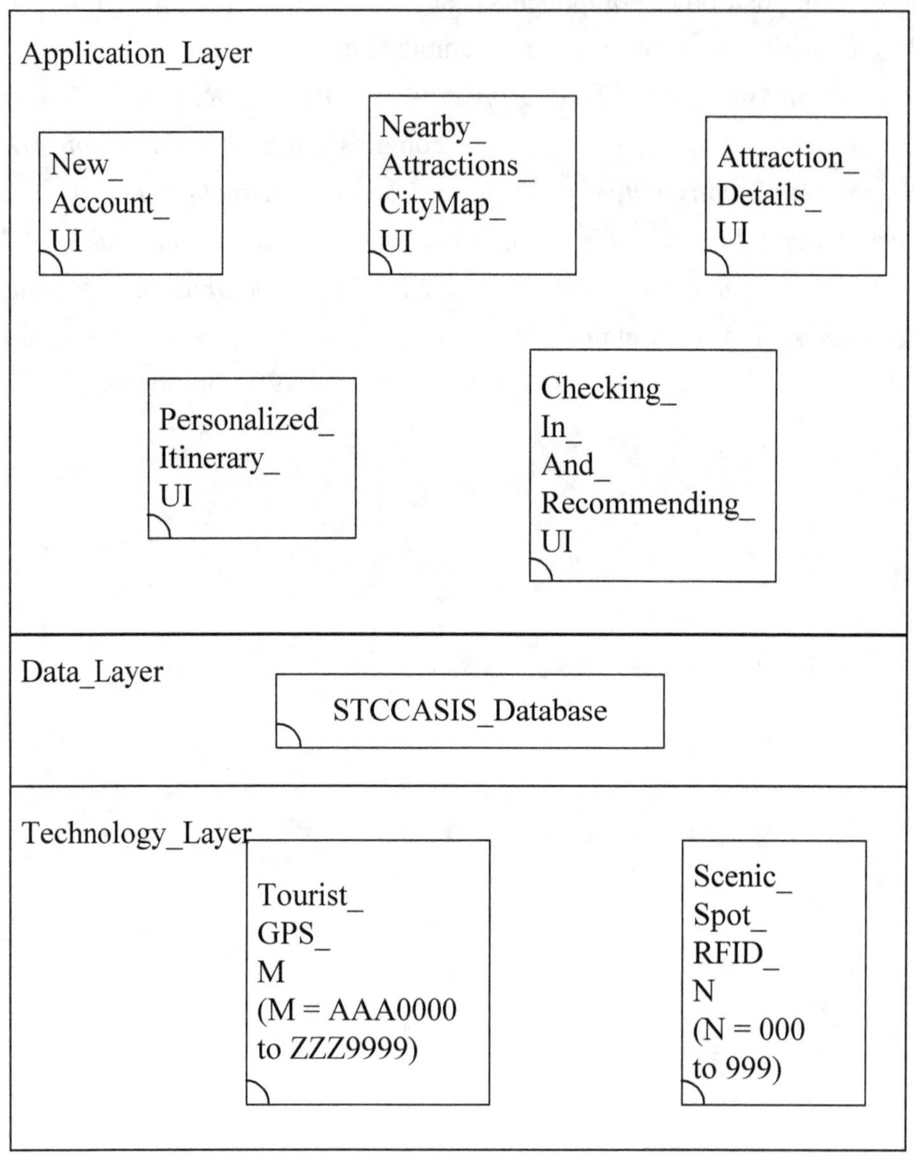

Figure 2-3 Structural Composition of
the *Smart Tourism City Cloud Applications and Services IoT System*

2-2 Behavior of Systems

Systems behavior refers to the interactions a system in conjunction with its environment. It is the response of a system to various stimuli, whether internal or external, conscious or subconscious, overt or covert, and voluntary or involuntary.

For example, Figure 2-4 demonstrates five individual behaviors: *Creating_New_Account*, *Showing_Nearby_Attractions_CityMap*, *Extracting_Attraction_Details*, *Planning_Personalized_Itinerary* and *Scenic_Spot_Checking_In_And_Recommending* that refer to the interactions the *Smart Tourism City Cloud Applications and Services IoT System* in conjunction with its environment.

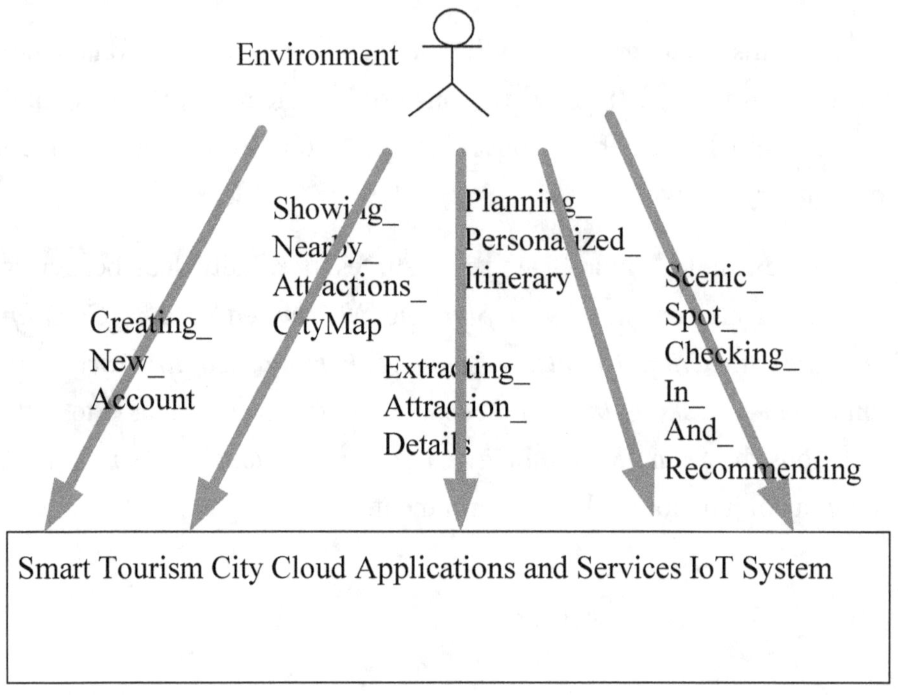

Figure 2-4 Behaviors of
the *Smart Tourism City Cloud Applications and Services IoT System*

For each behavior, the environment always initiates the interaction and will lead more follow-up interactions to be realized among components. For example, Figure 2-5 demonstrates that interactions between and among the environment and the *New_Account_UI* and *STCCASIS_Database* components shall draw forth the *Creating_New_Account* behavior.

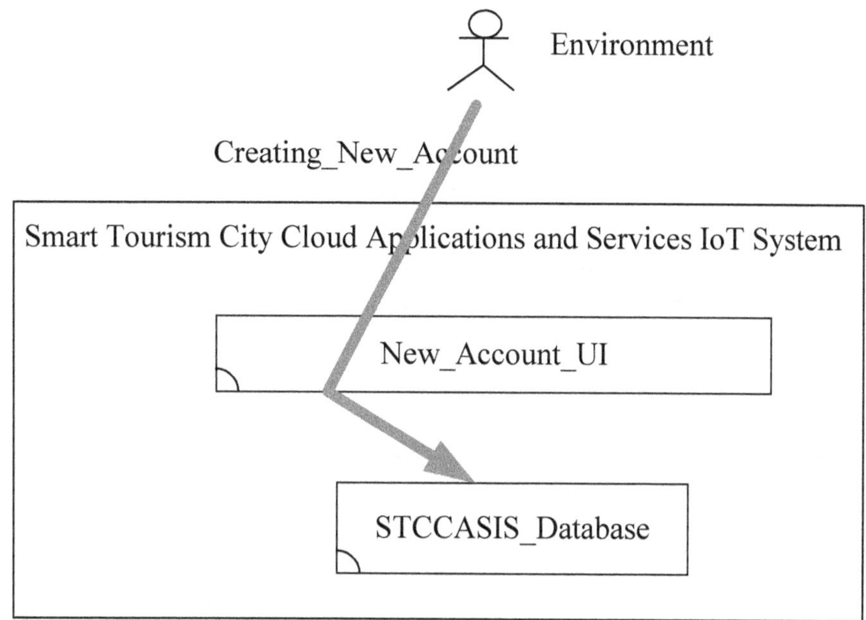

Figure 2-5 Interactions that Draw forth the *Creating_New_Account* Behavior

As a second example, Figure 2-6 demonstrates that interactions between and among the environment and the *Nearby_Attractions_CityMap_UI*, *STCCASIS_Database* and *Tourist_GPS_M (M = AAA0000 to ZZZ9999)* components shall draw forth the *Showing_Nearby_Attractions_CityMap* behavior.

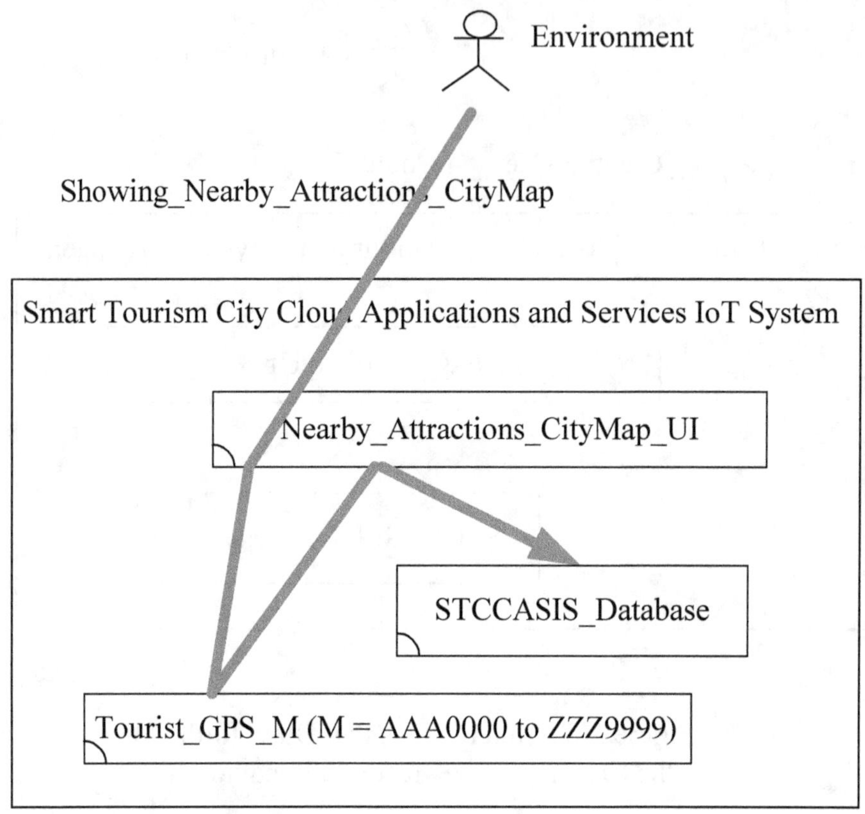

Figure 2-6 Interactions that Draw forth
the *Showing_Nearby_Attractions_CityMap* Behavior

As a third example, Figure 2-7 demonstrates that interactions between and among the environment and the *Attraction_Details_UI* and *STCCASIS_Database* components shall draw forth the *Extracting_Attraction_Details* behavior.

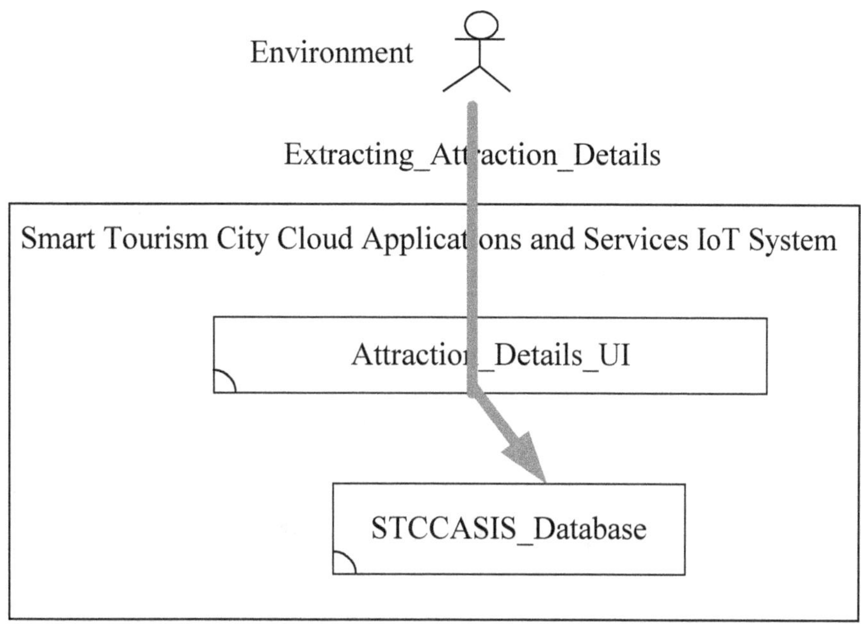

Figure 2-7 Interactions that Draw forth the *Extracting_Attraction_Details* Behavior

As a fourth example, Figure 2-8 demonstrates that interactions between and among the environment and the *Personalized_Itinerary_UI* and *STCCASIS_Database* components shall draw forth the *Planning_Personalized_Itinerary* behavior.

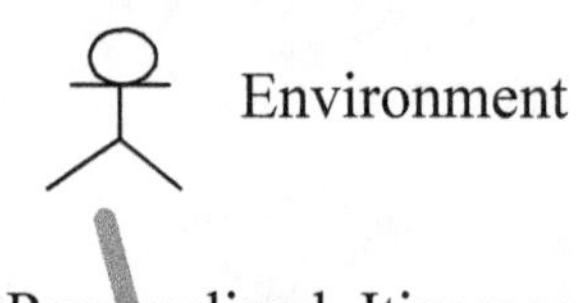

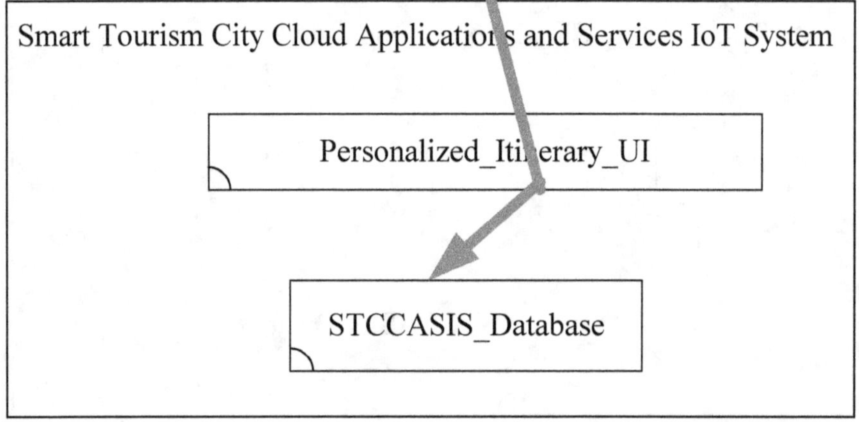

Figure 2-8 Interactions that Draw forth the *Planning_Personalized_Itinerary* Behavior

As a fifth example, Figure 2-9 demonstrates that interactions between and among the environment and the *Checking_In_And_Recommending_UI*, *STCCASIS_Database* and *Scenic_Spot_RFID_N (N = 000 to 999)* components shall draw forth the *Scenic_Spot_Checking_In_And_Recommending* behavior.

Scenic_Spot_Checking_In_And_Recommending

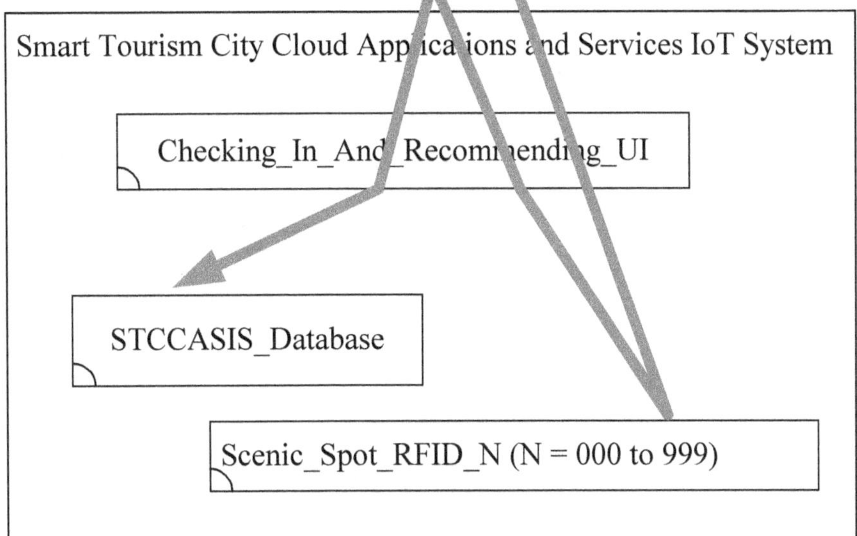

Figure 2-9 Interactions that Draw forth the *Scenic_Spot_Checking_In_And_Recommending* Behavior

Chapter 3: Structure-Behavior Coalescence

A system has been designed hopefully to be an integrated whole, embodied in its assembled components, their interactions with each other and the environment. Since systems structure and systems behavior are the two most prominent views of a system, integrating the systems structure and systems behavior apparently is the best way to achieve a truly integrated whole of a system. Because systems design 1.0 does not design the integration of systems structure and systems behavior, very likely it will never be able to actually form an integrated whole of a system.

Structure-behavior coalescence (SBC) provides an elegant way to integrate the systems structure and systems behavior, and hence achieves a truly integrated whole, of a system. A truly integrated whole sets a path to achieve the desired systems design. SBC facilitates an integrated whole. Therefore, we conclude that SBC sets a path to achieve the systems design. Systems design 2.0 uses the SBC approach and is highly adequate in designing a system.

3-1 Integrated Whole to Achieve the Systems Design

A system has been designed hopefully to be an integrated whole, embodied in its assembled components, their interactions with each other and the environment. In other words, an integrated whole sets a path to achieve the systems design as shown in Figure 3-1.

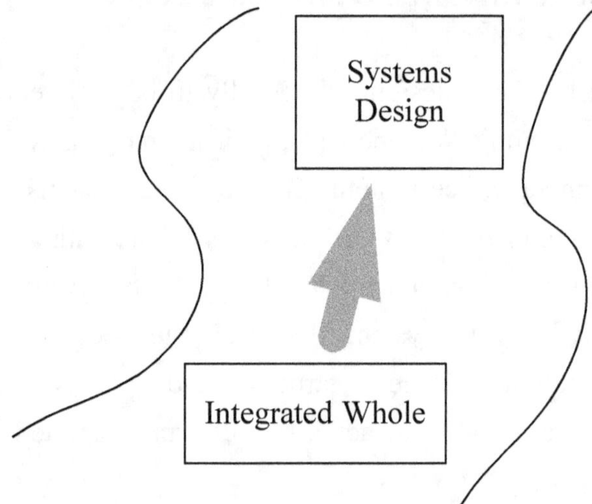

Figure 3-1 Integrated Whole to Achieve the Systems Design

In one systems design, different systems structures may draw forth the same integrated whole as shown in Figure 3-2.

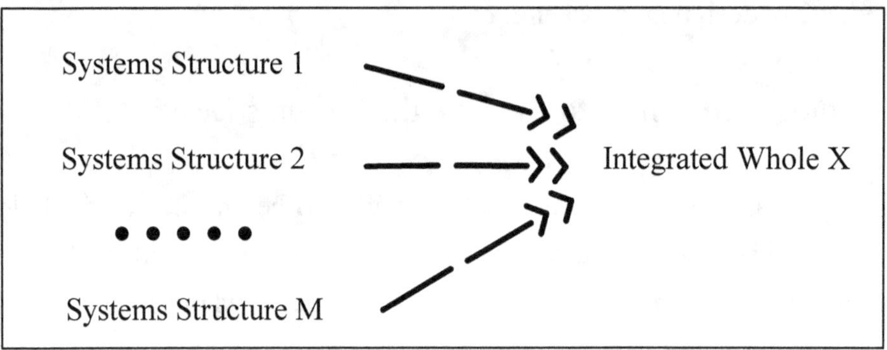

Figure 3-2 Different Systems Structures Draw Forth the Same Integrated Whole

Since there is only one systems structure exists in one systems design, one systems structure will draw forth one integrated whole as shown in Figure 3-3.

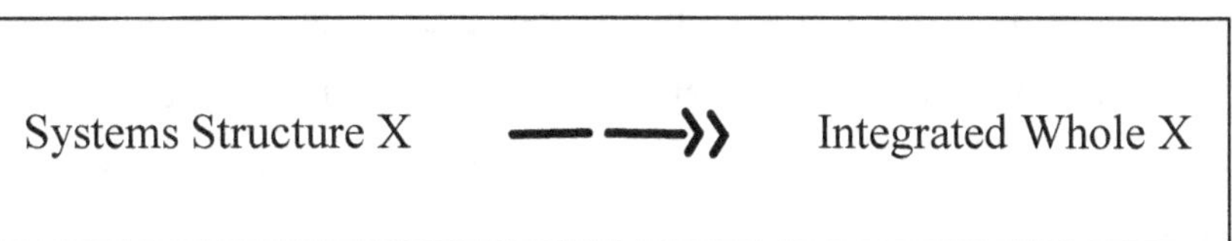

Figure 3-3 One Systems Structure Draws Forth One Integrated Whole

We conclude that in one systems design, an integrated whole must be attached to or built on a systems structure. In other words, an integrated whole shall not exist alone; it must be loaded on a systems structure just like a cargo is loaded on a ship as shown in Figure 3-4. There will be no integrated whole if there is no systems structure. A stand-alone integrated whole with no systems structure is not meaningful.

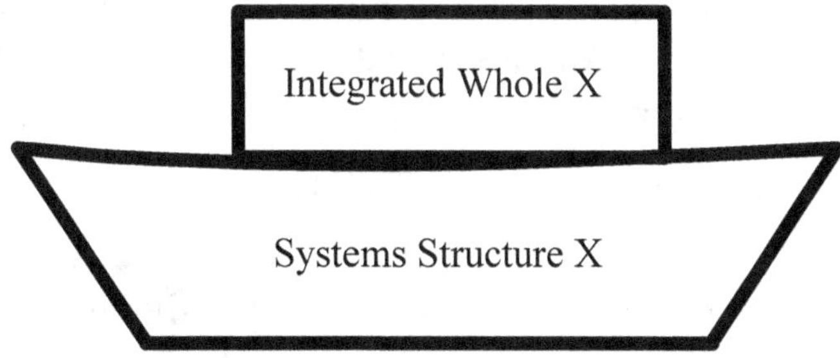

Figure 3-4 An Integrated Whole Must be Loaded on a Systems Structure

3-2 Integrating the Systems Structure and Systems Behavior

By integrating the systems structures and systems behaviors, we obtain structure-behavior coalescence (SBC) within a system. Since systems structures and systems behaviors are so tightly integrated, we sometimes claim that the core theme of structure-behavior coalescence is: "Systems Architecture = Systems Structure + Systems Behavior," as shown in Figure 3-5.

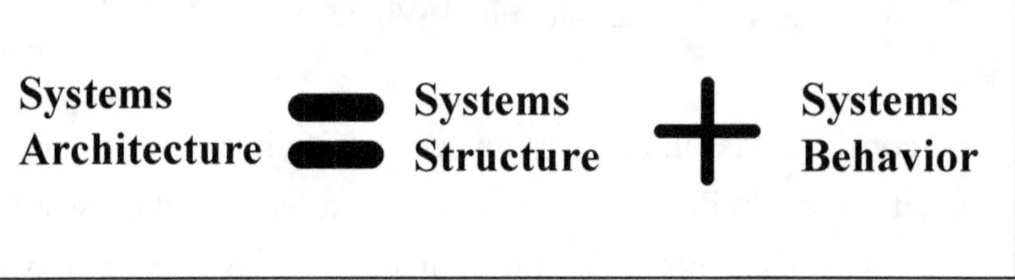

Figure 3-5 Core Theme of Structure-Behavior Coalescence

So far, integrating the systems structure and systems behavior has never been proposed or suggested besides the SBC approach. In most cases, systems behaviors are separated from systems structures when designing a system [Hoff10, Pres09, Shel11, Somm06].

3-3 Structure-Behavior Coalescence to Facilitate an Integrated Whole

Since systems structure and systems behavior are the two most prominent views of a system, integrating the systems structure and systems behavior apparently is the best way to achieve a truly integrated whole of a system. If we are not able to integrate the systems structure and systems behavior, then there is no way that we are able to integrate the whole system. In other words, structure-behavior coalescence (SBC) facilitates a truly integrated whole as shown in Figure 3-6.

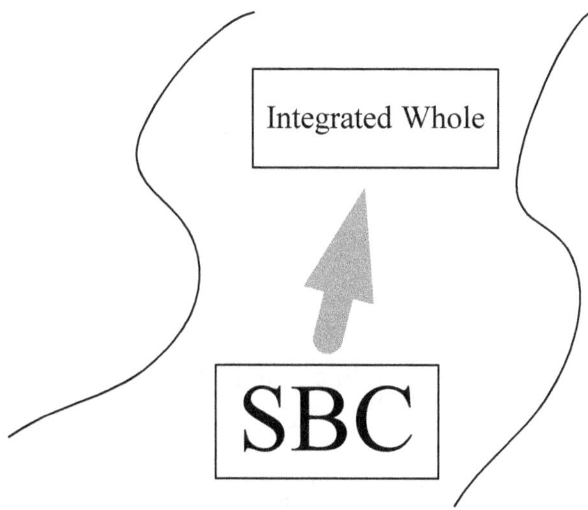

Figure 3-6 SBC Facilitates an Integrated Whole

Since systems design 1.0 does not design the integration of systems structure and systems behavior, very likely it will never be able to actually form an integrated whole of a system. In this situation, systems design 1.0 is powerless in designing a system adequately.

3-4 Structure-Behavior Coalescence to Achieve the System Design

Figure 3-1 declares that an integrated whole sets a path to achieve the desired systems design. Figure 3-6 declares that structure-behavior coalescence facilitates a truly integrated whole.

Combining the above two declarations, we conclude that the structure-behavior coalescence (SBC) approach sets a path to achieve the systems design as shown in Figure 3-7.

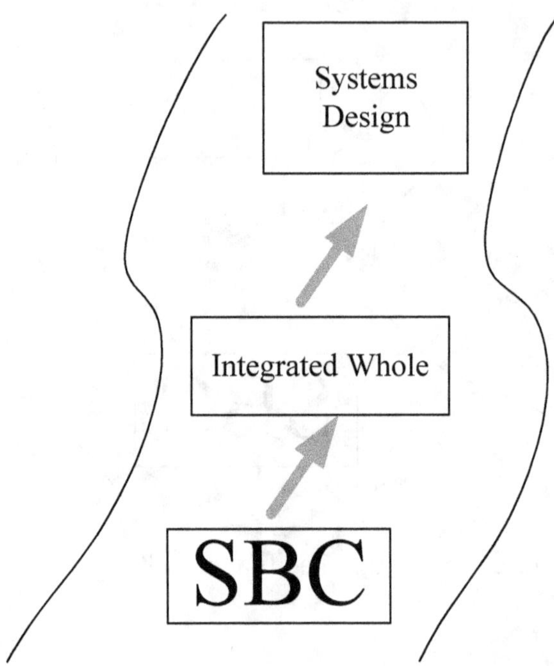

Figure 3-7 SBC to Achieve the Systems Design

In the SBC approach, different systems structures may draw forth the same systems behavior as shown in Figure 3-8.

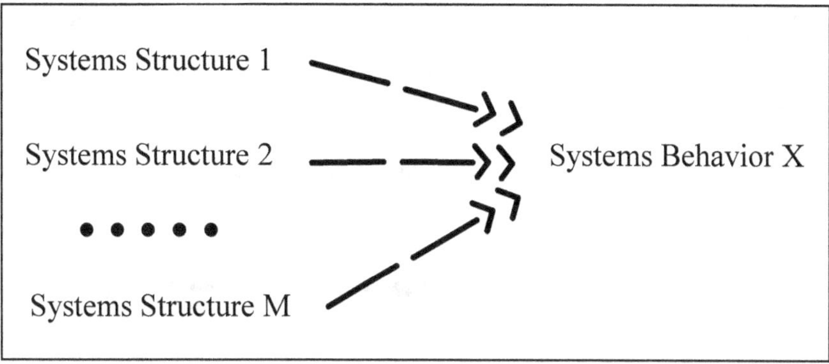

Figure 3-8 Different Systems Structures Draw Forth the Same Systems Behavior

Since there is only one systems structure exists in one systems design, one systems behavior will always be attached to or built on one systems structure as shown in Figure 3-9.

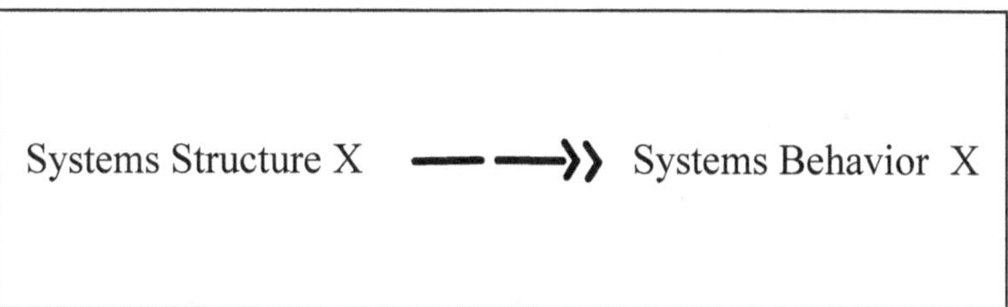

Figure 3-9 One Systems Behavior is Attached to One Systems Structure

We conclude that in the SBC approach, a systems behavior must be attached to or built on a systems structure. In other words, a systems behavior can not exist alone; it must be loaded on a systems structure just like a cargo is loaded on a ship as shown in Figure 3-10. There will be no systems behavior if there is no systems structure. A stand-alone systems behavior with no systems structure is not meaningful.

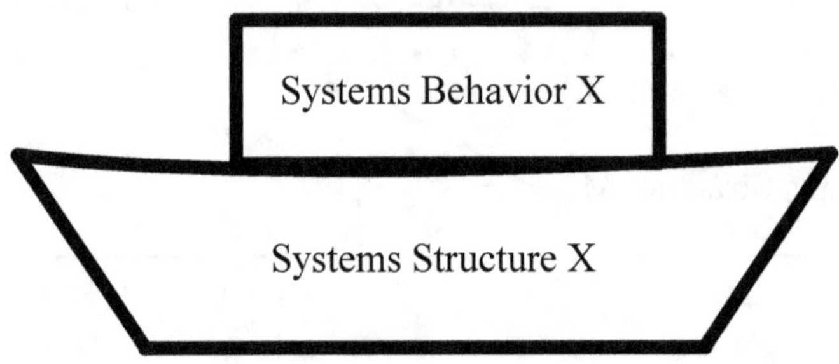

Figure 3-10 A Systems Behavior Must be Loaded on a Systems Structure

3-5 SBC Approach for Systems Design 2.0

Since structure-behavior coalescence (SBC) provides an elegant way to integrate the systems structure and systems behavior, we shall include it in the design of a system. Figure 3-11 shows how the systems design 2.0 designs a system.

> A system,
> through the SBC approach,
> truly is an integrated whole,
> embodied in its assembled components,
> their interactions with each other and the environment.

Figure 3-11 Systems Design 2.0 Designing a System

A system designed by the systems design 2.0 has the following characteristics: 1) it emphasizes the system's structure-behavior coalescence; 2) it is a truly integrated whole; 3) it is embodied in its assembled components; 4) components are interacting (or handshaking) [Chao15a, Chao15b, Chao15c, Chao15d, Chao15e, Hoar85, Miln89, Miln99] with each other and the environment; and 5) it uses structural decomposition [Chao14a, Chao14b, Chao14c, Ghar11] rather than functional decomposition [Scho10].

Structure-behavior coalescence (SBC) provides an elegant way to integrate the systems structure and systems behavior of a system. Systems design 2.0 uses the SBC approach to formally design the integration of systems structure and systems behavior of a system. Systems design 2.0 contains three fundamental diagrams: a) framework diagram, b) component operation diagram and c) interaction flow diagram.

So far, we have introduced the systems design 2.0 which should be able to appropriately design a system. In the following chapters, we shall elaborate the details of the systems design 2.0.

3-6 SBC Model Singularity

Channel-Based Single-Queue SBC Process Algebra (C-S-SBC-PA) [Chao17a], Channel-Based Multi-Queue SBC Process Algebra (C-M-SBC-PA) [Chao17b], Channel-Based Infinite-Queue SBC Process Algebra (C-I-SBC-PA) [Chao17c], Operation-Based Single-Queue SBC Process Algebra (O-S-SBC-PA) [Chao17d], Operation-Based Multi-Queue SBC Process Algebra (O-M-SBC-PA) [Chao17e] and Operation-Based Infinite-Queue SBC Process Algebra (O-I-SBC-PA) [Chao17f] are the six specialized SBC process algebras. The SBC process algebra (SBC-PA) shown in Figure 3-12 is a model singularity approach.

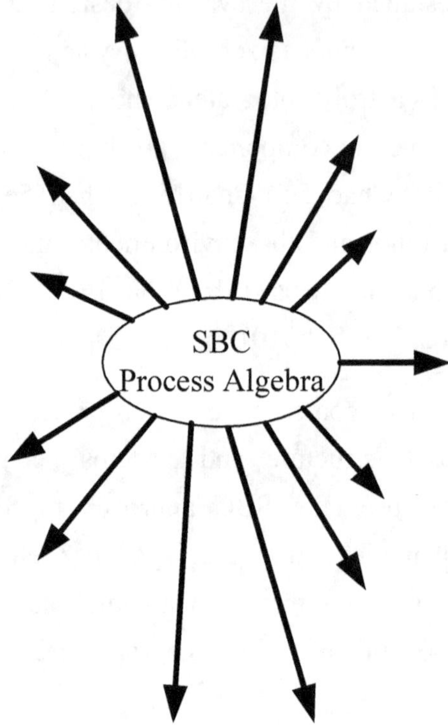

Figure 3-12 SBC-PA is a Model Singularity Approach.

The systems design 2.0 is also a model singularity approach. With SBC mind set sitting in the kernel, the systems design 2.0 single model shown in Figure 3-13 is therefore able to represent all structural views such as framework diagram (FD), component operation diagram (COD), and behavioral views such as interaction flow diagram (IFD).

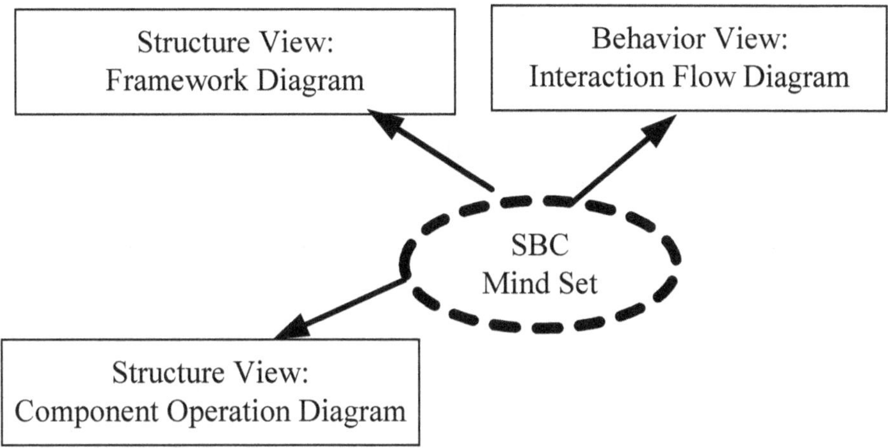

Figure 3-13 Systems Design 2.0 is a Model Singularity Approach.

The combination of SBC process algebra (SBC-PA) and systems design 2.0 is shown in Figure 3-14, again as a model singularity approach.

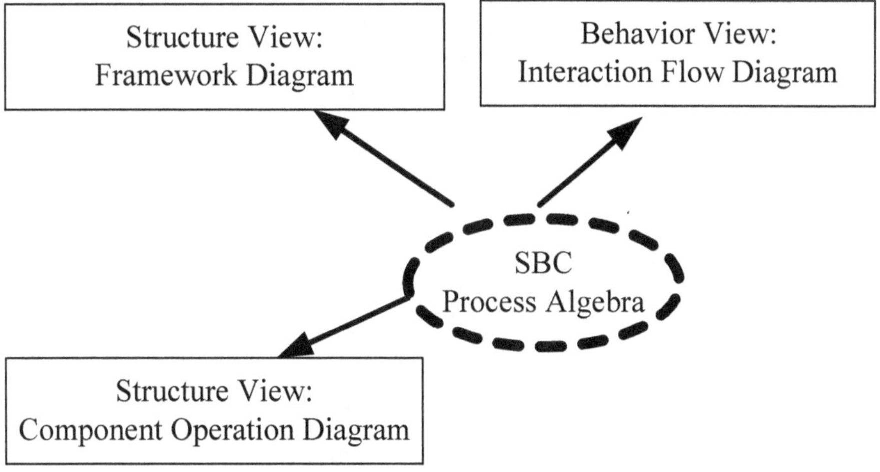

Figure 3-14 SBC Model is a Model Singularity Approach.

PART II: SBC APPROACH FOR SYSTEMS DESIGN 2.0

Chapter 4: Framework Diagram

SBC approach for systems design 2.0 uses a framework diagram (FD) to design the multi-layer (also referred to as multi-tier) decomposition and composition of a system.

4-1 Multi-Layer Decomposition and Composition

Decomposition and composition of a system can be designed in a multi-layer manner. We draw a framework diagram (FD) for the multi-layer decomposition and composition of a system.

As an example, Figure 4-1 shows a FD of the *Smart Tourism City Cloud Applications and Services IoT System* (STCCASIS). In the figure, *Application_Layer* contains the *New_Account_UI*, *Nearby_Attractions_CityMap_UI*, *Attraction_Details_UI*, *Personalized_Itinerary_UI* and *Checking_In_And_Recommending_UI* components; *Data_Layer* contains the *STCCASIS_Database* component; *Technology_Layer* contains the *Tourist_GPS_M (M = AAA0000 to ZZZ9999)* and *Scenic_Spot_RFID_N (N = 000 to 999)* components.

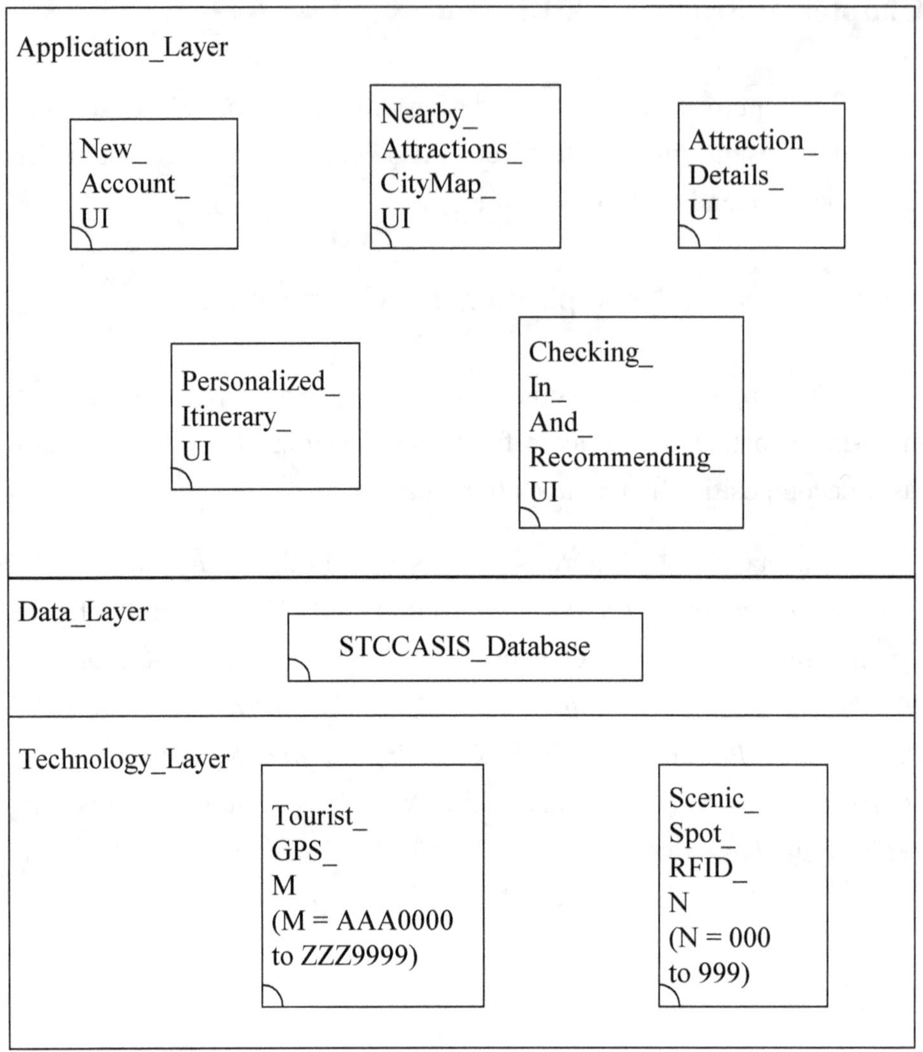

Figure 4-1 FD of the
Smart Tourism City Cloud Applications and Services IoT System

As the second example, Figure 4-2 shows a FD of the *Home Care Cloud Applications and Services IoT System* (HCCASIS). In the figure, *Presentation_Layer* and *Logic_Layer* are sub-layers of *Application_Layer*. *Presentation_Layer* contains the

Home_Account_Registering_UI, *Alerts_Notifying_UI*, *Emergency_Responses_UI* and *Monthly_Statistics_UI* components; *Logic_Layer* contains the *Sensor_Data_Acquisition_Daemon* component; *Data_Layer* contains the *HCCASIS_Database* component; *Technology_Layer* contains the *Position_Sensor_A00001*, *Position_Sensor_A00002,…*, *Position_Sensor_Z99999* components.

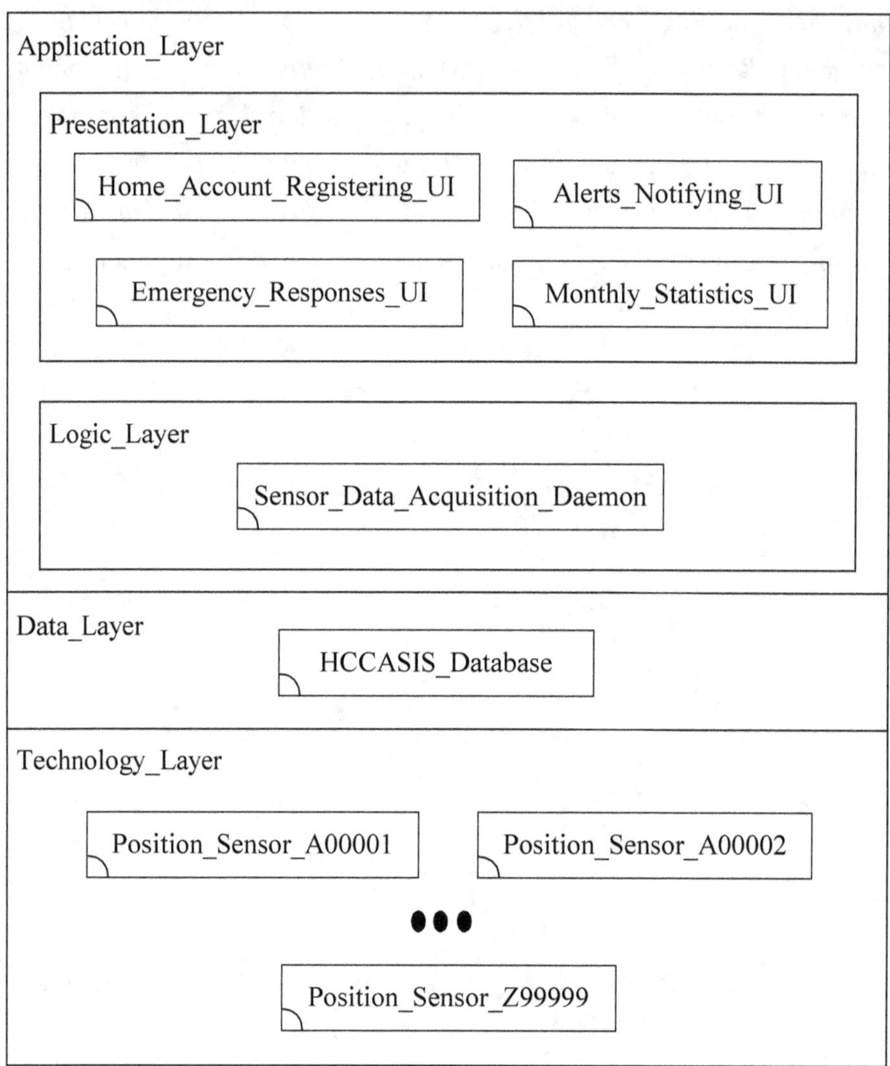

Figure 4-2 FD of the *Home Care Cloud Applications and Services IoT System*

4-2 Only Non-Aggregated Systems Appearing in Framework Diagrams

It is interesting that we see only non-aggregated systems shall appear in the multi-layer FD decomposition and composition of a system.

For the first example, Figure 4-1 in the previous section shows a FD of the *Smart Tourism City Cloud Applications and Services IoT System* (STCCASIS) in which only non-aggregated systems such as *New_Account_UI*, *Nearby_Attractions_CityMap_UI*, *Attraction_Details_UI*, *Personalized_Itinerary_UI*, *Checking_In_And_Recommending_UI*, *STCCASIS_Database*, *Tourist_GPS_M (M = AAA0000 to ZZZ9999)*, *Scenic_Spot_RFID_N (N = 000 to 999)* are displayed.

For a second example, Figure 4-2 in the previous section shows a FD of the *Home Care Cloud Applications and Services IoT System* (HCCASIS) in which only non-aggregated systems such as *Home_Account_Registering_UI*, *Alerts_Notifying_UI*, *Emergency_Responses_UI*, *Monthly_Statistics_UI*, *Sensor_Data_Acquisition_Daemon*, *HCCASIS_Database*, *Position_Sensor_A00001*, *Position_Sensor_A00002*,..., *Position_Sensor_Z99999* are displayed.

Chapter 5: Component Operation Diagram

SBC approach for systems design 2.0 uses a component operation diagram (COD) to design all components' operations of a system.

5-1 Operations of Each Component

An operation provided by each component represents a procedure or method or function of the component. If other components request this component to perform an operation, then shall use it to accomplish the operation request.

Each component in a system must possess at least one operation. A component should not exist in a system if it does not possess any operation. Figure 5-1 shows that the *Checking_In_And_Recommending_UI* component has two operations: *Scenic_Spot_Check_In* and *Scenic_Spot_Recommend.*

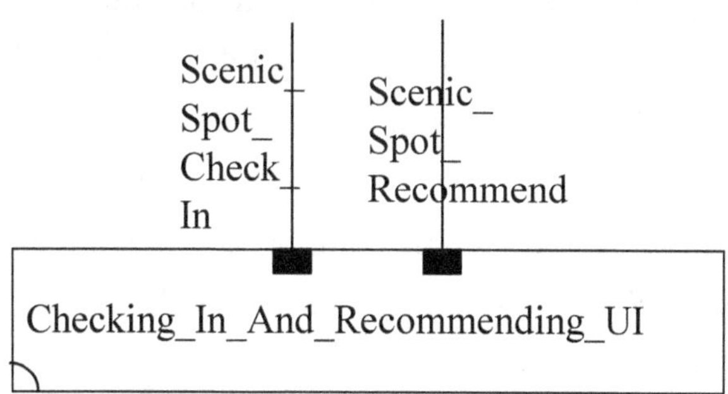

Figure 5-1 Two Operations of the *Checking_In_And_Recommending_UI* Component

An operation formula is utilized to fully represent an operation. An operation formula includes a) operation name, b) input parameters and c) output parameters as shown in Figure 5-2.

Operation_Name (In $i_1, i_2, \ldots, i_m$; Out $o_1, o_2, \ldots, o_n$)

Figure 5-2 Operation Formula

Operation name is the name of this operation. In a system, every operation name should be unique. Duplicate operation names shall not be allowed in any system.

An operation may have several input and output parameters. The input and output parameters, gathered from all operations, represent the input data and output data views of a system [Date03, Elma10]. As shown in Figure 5-3, component *Attraction_Details_UI* possesses the *Show_Attraction_Details* operation which has the *Scenic_Spot* input parameter (with the arrow direction pointing to the component) and the *Attraction_Details_Display* output parameter (with the arrow direction opposite to the component).

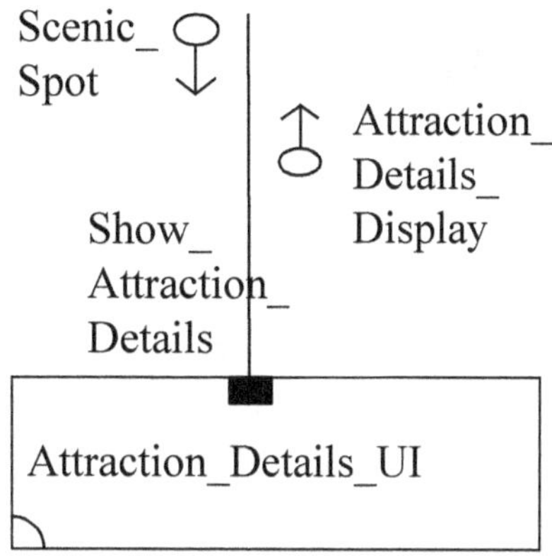

Figure 5-3 Input/Output Parameters

Data formats of input and output parameters can be described by data type specifications. There are two sets of data types: primitive and composite [Date03, Elma10]. Figure 5-4 shows the primitive data type specification of the *Scenic_Spot*, *Tourist_GPS_Coordinates* parameters occurring in the *Show_Attraction_Details(In Scenic_Spot; Out Attraction_Details_Display)*, *SQL_Select_Nearby_Attractions(In Tourist_GPS_Coordinates; Out Nearby_Attractions_Query)*, *SQL_Select_Attraction_Details(In Scenic_Spot; Out Attraction_Details_Query)*, *Tourist_GPS_Positioning(Out Tourist_GPS_Coordinates)* and *Scenic_Spot_RFID_Positioning(Out Scenic_Spot)* operation formulas.

Parameter	Data Type	Instances
Scenic_ Spot	Text	Vulcan Park and Museum
Tourist_ GPS_ Coordinates	Text	33.490565,-86.794727

Figure 5-4 Primitive Data Type Specification

Figure 5-5 shows the composite data type specification of the *Nearby_Attractions_Query* output parameter occurring in the *SQL_Select_Nearby_Attractions(In Tourist_GPS_Coordinates; Out Nearby_Attractions_Query)* operation formula.

<table>
<tr><td>Parameter</td><td>Nearby_Attractions_Query</td></tr>
<tr><td>Data Type</td><td>TABLE of
Scenic_Spot: Text
Scenic_Spot_GPS_Coordinates: Text
End TABLE ;</td></tr>
<tr><td>Instances</td><td>
<table>
<tr><th>Scenic_Spot</th><th>Scenic_Spot_ GPS_Coordinates</th></tr>
<tr><td>Birmingham Zoo</td><td>33.48657862, -86.77911758</td></tr>
<tr><td>Vulcan Park and Museum</td><td>33.490565, -86.794727</td></tr>
<tr><td>McWane Science Center</td><td>33.51520752, -86.80830002</td></tr>
<tr><td>Birmingham's Railroad Park</td><td>33.5099169, -86.8084382</td></tr>
<tr><td>University of Alabama at Birmingham</td><td>33.49302095, -86.80898666</td></tr>
</table>
</td></tr>
</table>

Figure 5-5 Composite Data Type Specification of *Nearby_Attractions_Query*

Figure 5-6 shows the composite data type specification of the *Attraction_Details_Query* output parameter occurring in the *SQL_Select_Attraction_Details(In Scenic_Spot; Out Attraction_Details_Query)* operation formula.

<table>
<tr><td>Parameter</td><td>Attraction_Details_Query</td></tr>
<tr><td>Data Type</td><td>TABLE of
  Scenic_Spot: Text
  Scenic_Spot_Address: Text
  Description: Text
  Main_Image: ImageData
End TABLE ;</td></tr>
<tr><td>Instances</td><td>
<table>
<tr><th>Scenic_Spot</th></tr>
<tr><td>Vulcan Park and Museum</td></tr>
</table>
<table>
<tr><th>Scenic_Spot_Address</th></tr>
<tr><td>1701 Valley View Dr,
Birmingham, AL 35209,
U.S.A.</td></tr>
</table>
<table>
<tr><th>Description</th><th>Main_Image</th></tr>
<tr><td>Vulcan is the world’s largest cast iron statue; made of 100,000 pounds of iron and 56 feet tall, he stands at the top of Red Mountain overlooking the city of Birmingham. But Vulcan is more than just a statue—Vulcan Park and Museum features spectacular views of Birmingham, an interactive history museum that examines Vulcan and Birmingham’s story, a premier venue for private events, and a beautiful public park for visitors and locals to enjoy. With an official information center operated by the Greater Birmingham Convention and Visitors Bureau, Vulcan Park and Museum serves as the first stop for visitors to the Birmingham area!
-- Courtesy of visitvulcan.com --</td><td>011101000010101010
010101001000000010
010010101001010010
101010010101001010
010010100101110100
001010101001010100
100000001001001010
100100101010010100
101010100101010010
100100101001011101
000010101010010101
001000000010010010
101001010010101010
010101001010010010
100101110100001010
101001010100100000</td></tr>
</table>
</td></tr>
</table>

Figure 5-6 Composite Data Type Specification of *Attraction_Details_Query*

5-2 Drawing the Component Operation Diagram

For a system, COD is used to design all components’ operations. Figure 5-7 shows the *Smart Tourism City Cloud Applications and Services IoT System’s COD*. In the figure, component *New_Account_UI* has one operation: *Input_New_Account*; component *Nearby_Attractions_CityMap_UI* has one operation: *Show_Nearby_Attractions_CityMap*; component *Attraction_Details_UI* has one operation: *Show_Attraction_Details*; component *Personalized_Itinerary_UI* has one operation: *Input_Personalized_Itinerary*; component *Checking_In_And_Recommending_UI* has two operations: *Scenic_Spot_Check_In* and *Scenic_Spot_Recommend*; component *STCCASIS_Database* has five operations: *SQL_Insert_New_Account*, *SQL_Select_Nearby_Attractions*, *SQL_Select_Attraction_Details*, *SQL_Insert_Personalized_Itinerary* and *SQL_Insert_Checking_In_And_Recommending*; component *Tourist_GPS_M (M = AAA0000 to ZZZ9999)* has one operation: *Tourist_GPS_Positioning*; component *Scenic_Spot_RFID_N (N = 000 to 999)* has one operation: *Scenic_Spot_RFID_Positioning*.

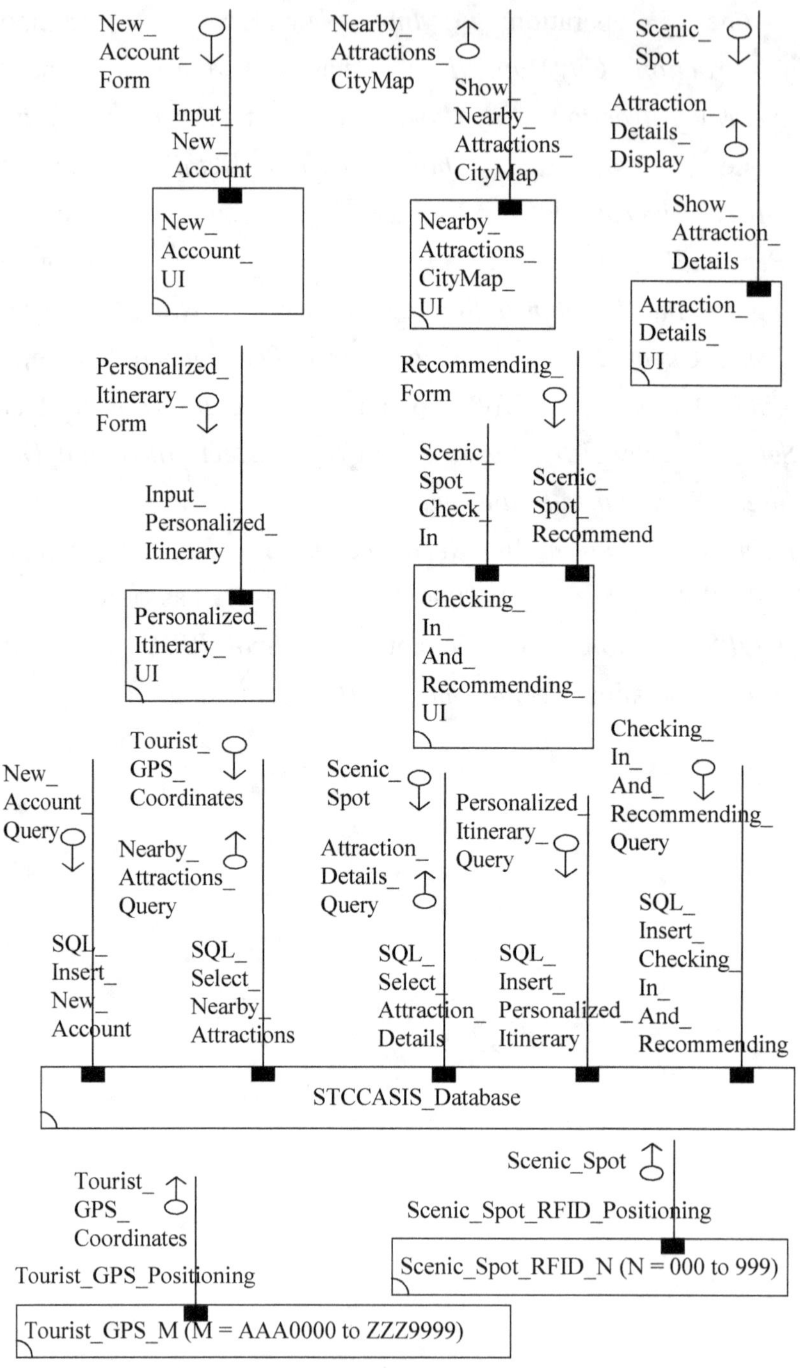

Figure 5-7 COD of the
Smart Tourism City Cloud Applications and Services IoT System

Chapter 6: Interaction Flow Diagram

SBC approach for systems design 2.0 uses an interaction flow diagram (IFD) to design each individual behavior of the overall behavior of a system.

6-1 Individual Behavior Represented by Interaction Flow Diagram

The overall behavior of a system consists of many individual behaviors. Each individual behavior represents an execution path. An IFD is utilized to design such an individual behavior.

Figure 6-1 demonstrates that the *Smart Tourism City Cloud Applications and Services IoT System* (STCCASIS) has five behaviors; thus, it has five IFDs.

System	IFD
STCCASIS	Creating_New_Account
	Showing_Nearby_Attractions_CityMap
	Extracting_Attraction_Details
	Planning_Personalized_Itinerary
	Scenic_Spot_Checking_In_And_Recommending

Figure 6-1 *Smart Tourism City Cloud Applications and Services IoT System* has Five IFDs

Figure 6-2 demonstrates that the *Home Care Cloud Applications and Services IoT System* (HCCASIS) has five behaviors; thus, it has five IFDs.

System	IFD
HCCASIS	Registering_Home_Account
	Recording_Occurring_Landslide
	Alerts_Notifying
	Recording_Emergency_Responses
	Printing_Monthly_Statistics

Figure 6-2 *Home Care Cloud Applications and Services IoT System* has Five IFDs

6-2 Drawing the Interaction Flow Diagram

Let us now explain the usage of interaction flow diagram (IFD) by drawing an IFD step by step. Figure 6-3 demonstrates an IFD of the *Planning_Personalized_Itinerary* behavior. The X-axis direction is from the left side to right side and the Y-axis direction is from the above to the below. Inside an IFD, there are four elements: a) external environment's actor, b) components, c) interactions and d) input/output parameters. Participants of the interaction, such as the external environment's actor and each component, are laid aside along the X-axis direction on the top of the diagram. The external environment's actor which initiates the sequential interactions is always placed on the most left side of the X-axis. Then, interactions among the external environment's actor and components successively in turn decorate along the Y-axis direction. The first interaction is placed on the top of the Y-axis position. The last interaction is placed on the bottom of the Y-axis position. Each interaction may carry several input and/or output parameters.

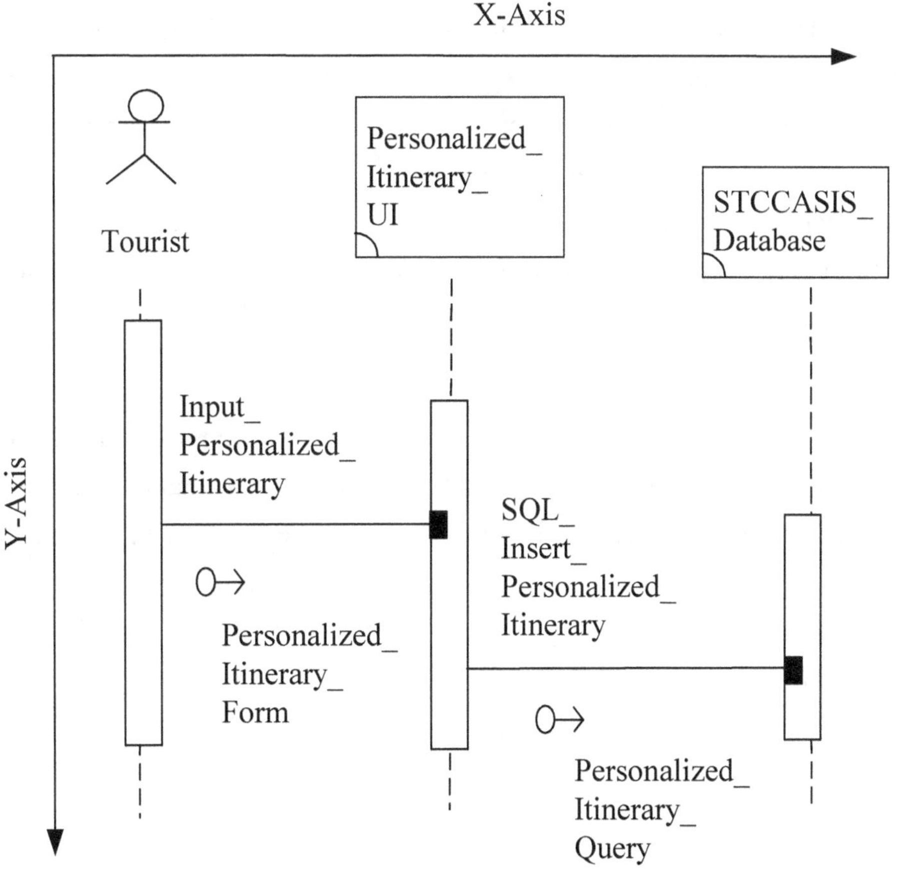

Figure 6-3 IFD of the *Planning_Personalized_Itinerary* Behavior

In Figure 6-3, *Tourist* is an external environment's actor. *Personalized_Itinerary_UI* and *STCCASIS_Database* are components. *Input_Personalized_Itinerary* is an operation, carrying the *Personalized_Itinerary_Form* input parameter, which is provided by the *Personalized_Itinerary_UI* component. *SQL_Insert_Personalized_Itinerary* is an operation, carrying the *Personalized_Itinerary_Query* input parameter, which is provided by the *STCCASIS_Database* component.

The execution path of Figure 6-3 is as follows. First, actor *Tourist* interacts with the *Personalized_Itinerary_UI* component through the *Input_Personalized_Itinerary* operation call interaction, carrying the *Personalized_Itinerary_Form* input parameter. Finally, component *Personalized_Itinerary_UI* interacts with the *STCCASIS_Database* component through the *SQL_Insert_Personalized_Itinerary* operation call interaction, carrying the *Personalized_Itinerary_Query* input parameter.

For each interaction, the solid line stands for operation call while the dashed line stands for operation return. The operation call and operation return interactions, if using the same operation name, belong to the identical operation. Figure 6-4 exhibits two interactions (operation call interaction and operation return interaction) having the identical "*Show_Attraction_Details*" operation.

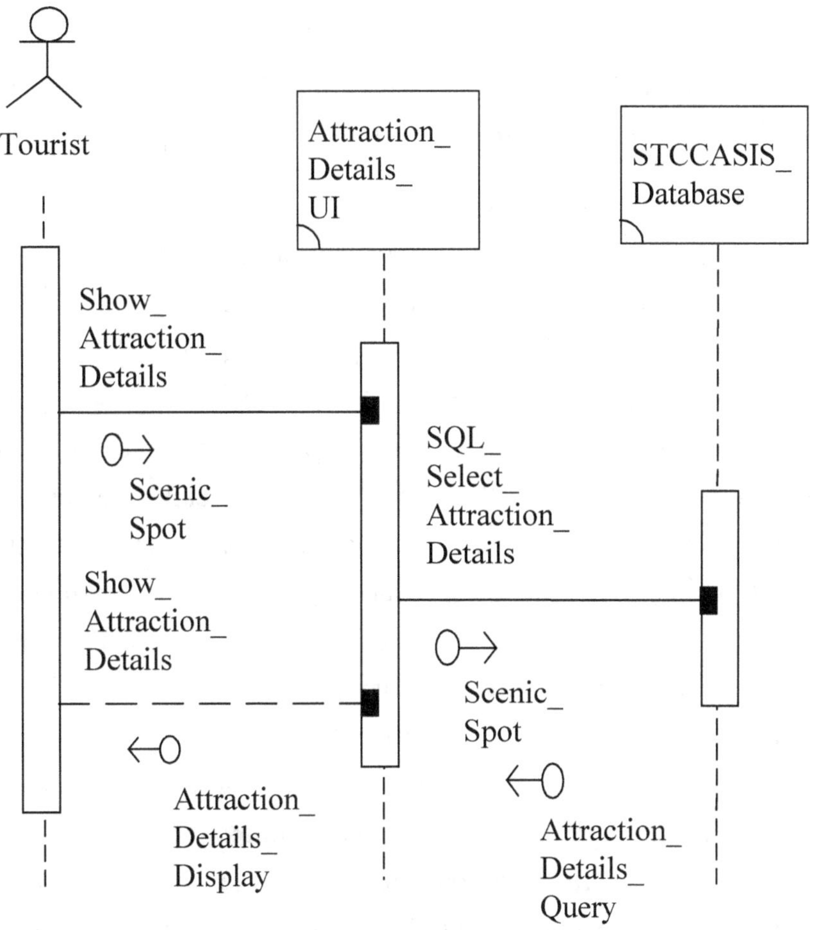

Figure 6-4 Two Interactions Have the Identical Operation

The execution path of Figure 6-4 is as follows. First, actor *Tourist* interacts with the *Attraction_Details_UI* component through the *Show_Attraction_Details* operation call interaction, carrying the *Scenic_Spot* input parameter. Next, component *Attraction_Details_UI* interacts with the *STCCASIS_Database* component through the *SQL_Select_Attraction_Details* operation call interaction, carrying the

Scenic_Spot input parameter and *Attraction_Details_Query* output parameter. Finally, actor *Tourist* interacts with the *Attraction_Details_UI* component through the *Show_Attraction_Details* operation return interaction, carrying the *Attraction_Details_Display* output parameter.

An interaction flow diagram may contain a conditional expression. Figure 6-5 shows such an example which has the following execution path. First, external environment's actor *Employee* interacts with the *Computer* component through the *Open* operation call interaction, carrying the *Task_No* input parameter. Next, if the *var_1 < 4 & var_2 > 7* condition is true then component *Computer* shall interact with the *Skype* component through the *Op_1* operation call interaction and component *Skype* shall interact with the *Earphone* component through the *Op_4* operation call interaction, carrying the *Skype_Earphone* output parameter; else if the *var_3 = 99* condition is true then component *Computer* shall interact with the *Skype* component through the *Op_2* operation call interaction and component *Skype* shall interact with the *Speaker* component through the *Op_5* operation call interaction, carrying the *Skype_Speaker* output parameter; else component *Computer* shall interact with the *Youtube* component through the *Op_3* operation call interaction and component *Youtube* shall interact with the *Speaker* component through the *Op_6* operation call interaction, carrying the *Youtube_Speaker* output parameter. Continuingly, if the *var_1 < 4 & var_2 > 7* condition is true then component *Computer* shall interact with the *Skype* component through the *Op_1* operation return interaction, carrying the *Status_1* output parameter; else if the *var_3 = 99* condition is true then component *Computer* shall interact with the *Skype* component through the *Op_2* operation return interaction, carrying the *Status_2* output parameter; else component *Computer* shall interact with the *Youtube* component through the *Op_3* operation return interaction, carrying the *Status_3* output parameter. Finally, external environment's actor *Employee* interacts with the *Computer* component through the *Open* operation return interaction, carrying the *Status* output parameter.

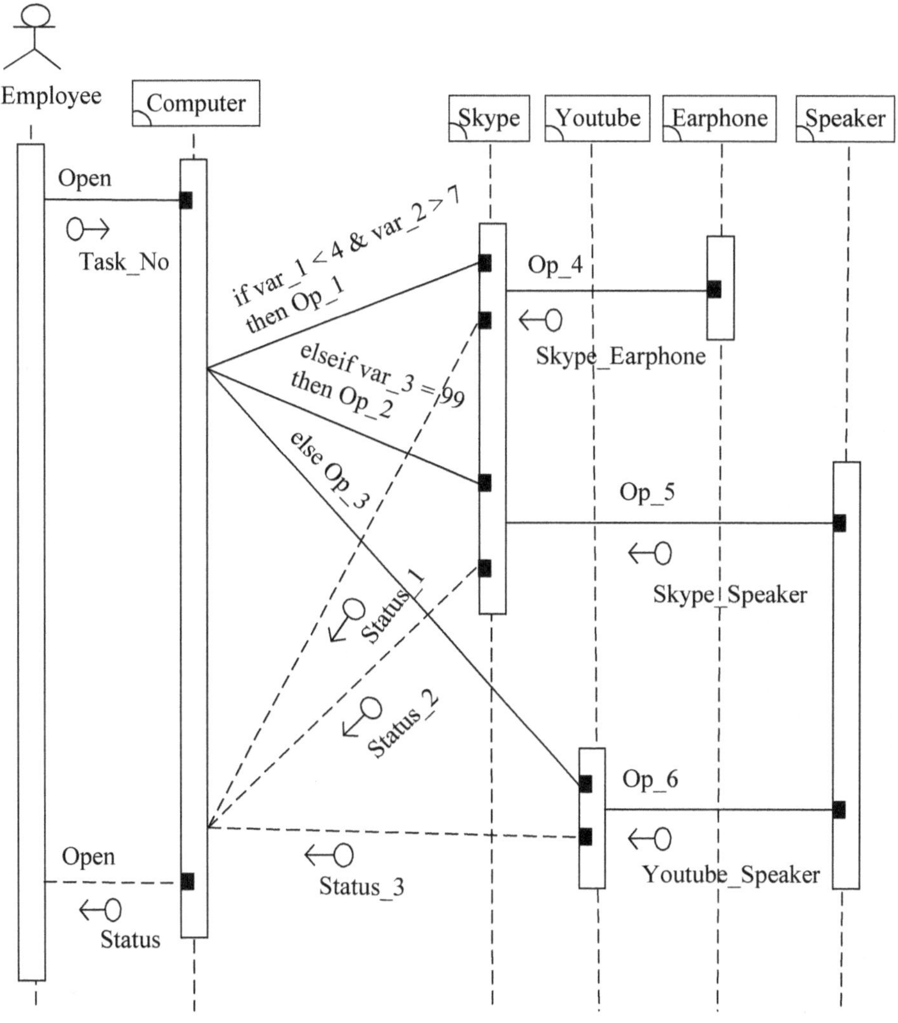

Figure 6-5 Conditional Interaction

Several Boolean conditions are shown in Figure 6-5. They are "*var_1 < 4 & var_2 > 7*" and "*var_3 = 99*". Variables, such as *var_1*, *var_2* and *var_3*, appearing in the Boolean condition can be local or global variables [Prat00, Seth96].

PART III: CASES STUDY

Chapter 7: Systems Design 2.0 of the Smart Tourism City Cloud Applications and Services IoT System

The smart tourism cities [Buha00] concept emerges from the development of smart cities [Pask09]. With cloud Internet of Things (IoT) technology being embedded on all organizations and entities, tourism cities will exploit synergies between ubiquitous sensing technology and their social networks to support the enrichment of tourist experiences. The smart cities strategy is an inevitable trend in the future development of all cities around the world. The smart tourism city is an important part and a practical attempt of the smart city strategy. This smart city strategy attempts to combine the cloud IoT technology with the development of the smart tourism industry and smart tourism cities.

Advancements in cloud applications and services IoT systems present enormous potential for accurate monitoring and providing service to the customers and administrators. A cost effective solution to this service can be provided by wireless sensor networks which consists of larger number of sensors placed on the existing land areas without installing expensive cabling and are capable of adjusting with the easily available sensors. The information obtained from each sensor is processed collaboratively to evaluate meaning metrics such as landslide signs of all land areas. A smart tourism city cloud applications and services IoT system (STCCASIS) offers a lot of integrated opportunities in urban city areas that are expected to support sustainable economic development, high quality of life, and foster participation and engagement of citizens.

STCCASIS helps tourists capture the moment emotion and process all these data around users, and turn them into not just helpful information, or even personalized knowledge. Behaviors of STCCASIS consist of: a) behavior of *Creating_New_Account*, b) behavior of

Showing_Nearby_Attractions_CityMap, c) behavior of *Extracting_Attraction_Details*, d) behavior of *Planning_Personalized_Itinerary* and e) behavior of *Scenic_Spot_Checking_In_And_Recommending.*

Using the structure-behavior coalescence (SBC) approach, we shall go through: a) framework diagram, b) component operation diagram and c) interaction flow diagram, to accomplish the systems design 2.0 for STCCASIS.

7-1 Framework Diagram

Systems design 2.0 uses a framework diagram (FD) to design the multi-layer composition and decomposition of the *Smart Tourism City Cloud Applications and Services IoT System* (STCCASIS) as shown in Figure 7-1.

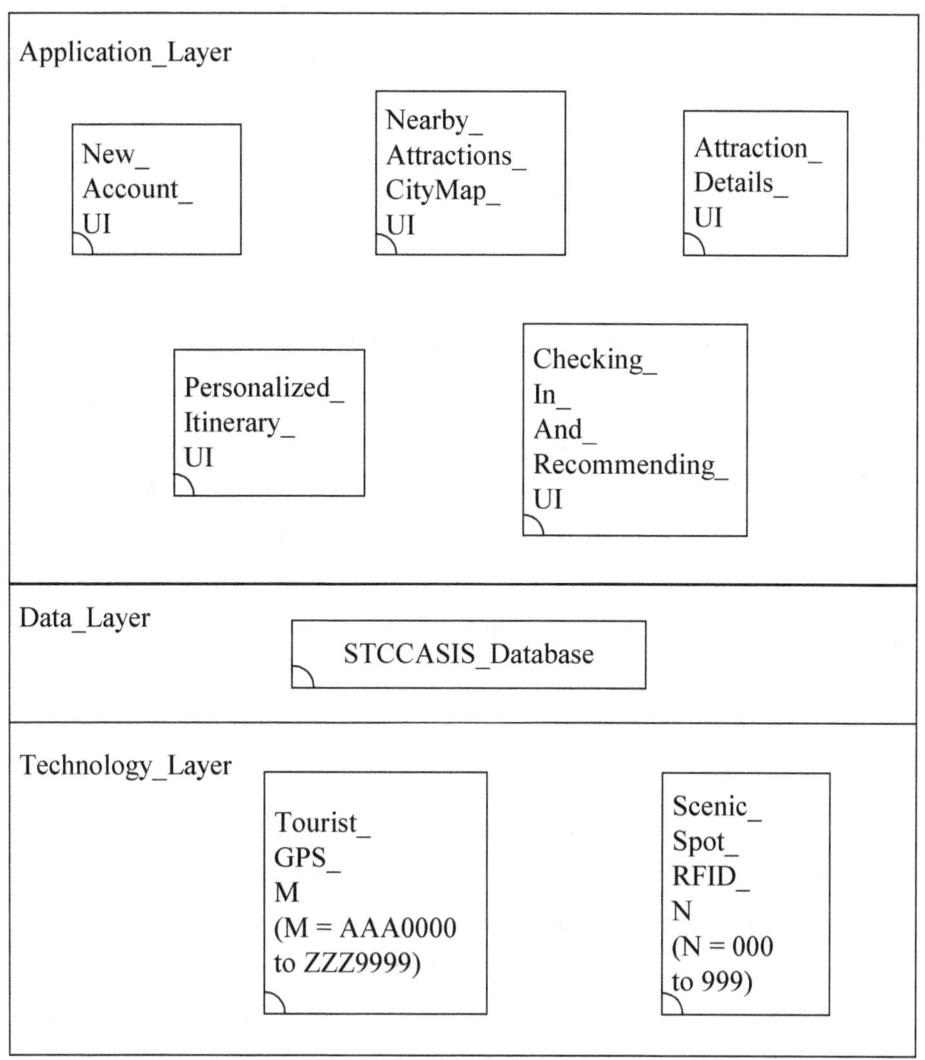

Figure 7-1 FD of the *STCCASIS*

In the above figure, *Application_Layer* contains the *New_Account_UI*, *Nearby_Attractions_CityMap_UI*, *Attraction_Details_UI*, *Personalized_Itinerary_UI* and *Checking_In_And_Recommending_UI* components; *Data_Layer* contains the *STCCASIS_Database* component; *Technology_Layer* contains the *Tourist_GPS_M (M = AAA0000 to ZZZ9999)* and *Scenic_Spot_RFID_N (N = 000 to 999)* components.

7-2 Component Operation Diagram

Systems design 2.0 uses a component operation diagram (COD) to design the operations of all components of the *Smart Tourism City Cloud Applications and Services IoT System* (STCCASIS) as shown in Figure 7-2. In the figure, component *New_Account_UI* has one operation: *Input_New_Account*; component *Nearby_Attractions_CityMap_UI* has one operation: *Show_Nearby_Attractions_CityMap*; component *Attraction_Details_UI* has one operation: *Show_Attraction_Details*; component *Personalized_Itinerary_UI* has one operation: *Input_Personalized_Itinerary*; component *Checking_In_And_Recommending_UI* has two operations: *Scenic_Spot_Check_In* and *Scenic_Spot_Recommend*; component *STCCASIS_Database* has five operations: *SQL_Insert_New_Account*, *SQL_Select_Nearby_Attractions*, *SQL_Select_Attraction_Details*, *SQL_Insert_Personalized_Itinerary* and *SQL_Insert_Checking_In_And_Recommending*; component *Tourist_GPS_M (M = AAA0000 to ZZZ9999)* has one operation: *Tourist_GPS_Positioning*; component *Scenic_Spot_RFID_N (N = 000 to 999)* has one operation: *Scenic_Spot_RFID_Positioning*.

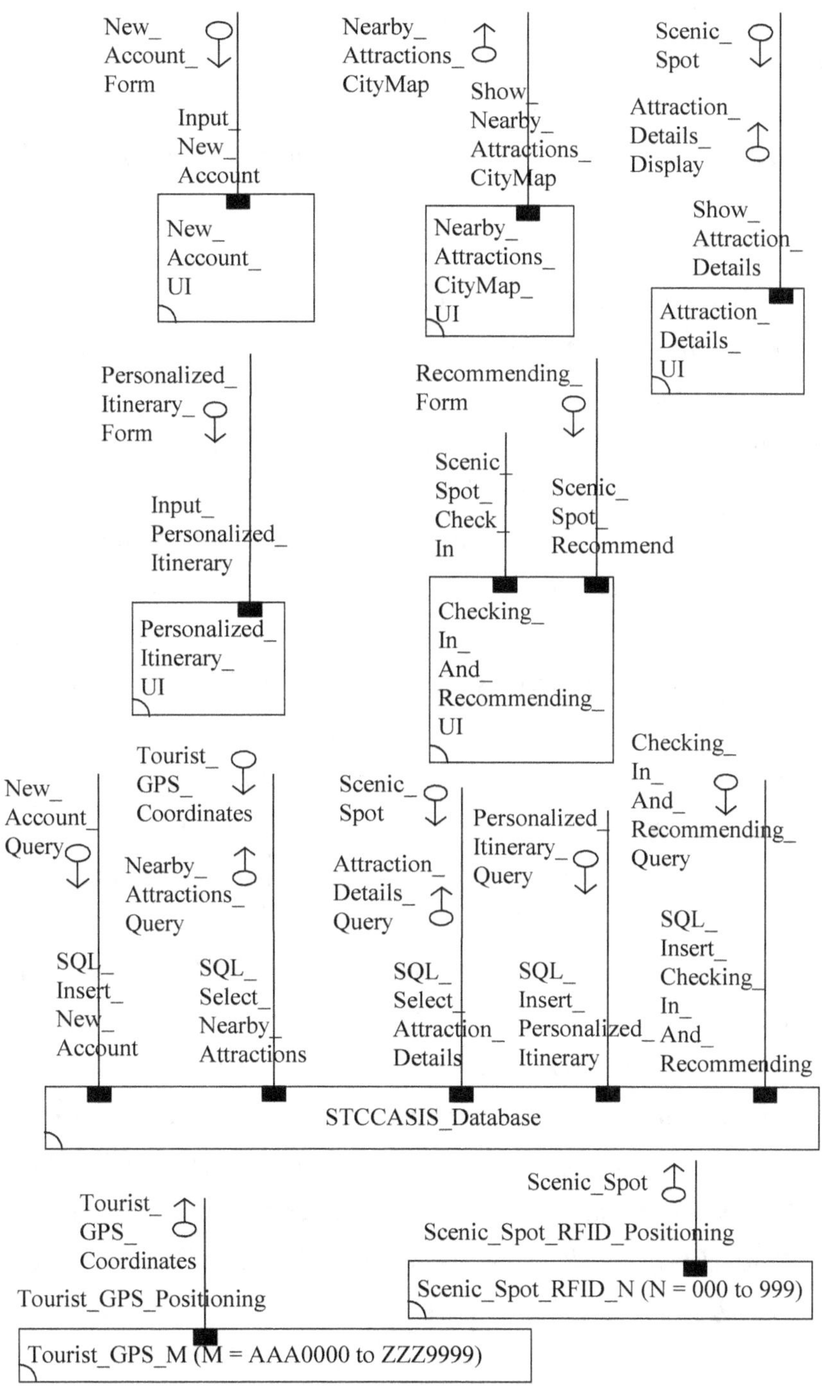

Figure 7-2 COD of the *STCCASIS*

The operation formula of *Input_New_Account* is *Input_New_Account(In New_Account_Form)*. The operation formula of *Show_Nearby_Attractions_CityMap* is *Show_Nearby_Attractions_CityMap(Out Nearby_Attractions_CityMap)*. The operation formula of *Show_Attraction_Details* is *Show_Attraction_Details(In Scenic_Spot; Out Attraction_Details_Display)*. The operation formula of *Input_Personalized_Itinerary* is *Input_Personalized_Itinerary(In Personalized_Itinerary_Form)*. The operation formula of *Scenic_Spot_Check_In* is *Scenic_Spot_Check_In*. The operation formula of *Scenic_Spot_Recommend* is *Scenic_Spot_Recommend(In Recommending_Form)*. The operation formula of *SQL_Insert_New_Account* is *SQL_Insert_New_Account(In New_Account_Query)*. The operation formula of *SQL_Select_Nearby_Attractions* is *SQL_Select_Nearby_Attractions(In Tourist_GPS_Coordinates; Out Nearby_Attractions_Query)*. The operation formula of *SQL_Select_Attraction_Details* is *SQL_Select_Attraction_Details(In Scenic_Spot; Out Attraction_Details_Query)*. The operation formula of *SQL_Insert_Personalized_Itinerary* is *SQL_Insert_Personalized_Itinerary(In Personalized_Itinerary_Query)*. The operation formula of *SQL_Insert_Checking_In_And_Recommending* is *SQL_Insert_Checking_In_And_Recommending(In Checking_In_And_Recommending_Query)*. The operation formula of *Tourist_GPS_Positioning* is *Tourist_GPS_Positioning(Out Tourist_GPS_Coordinates)*. The operation formula of *Scenic_Spot_RFID_Positioning* is *Scenic_Spot_RFID_Positioning(Out Scenic_Spot)*.

Figure 7-3 shows the composite data type specification of the *New_Account_Form* input parameter occurring in the *Input_New_Account(In New_Account_Form)* operation formula.

Parameter	*New_Account_Form*
Data Type	TABLE of Username: Text Email_Address: Text First_Name: Text Last_Name: Text Address: Text City: Text State: Text Country: Text End TABLE ;
Instances	STCCASIS Smart Tourism City Birminghamn **New Account Form** Username: A1B2C3D4 Email_Address: edgar6789@gmail.com First_Name: Edgar Last_Name: Ashworth Address: 702 Ross Street City: Dallas State: Texas Country: U.S.A.

Figure 7-3 Composite Data Type Specification of *New_Account_Form*

Figure 7-4 shows the composite data type specification of the *Nearby_Attractions_CityMap* output parameter occurring in the *Show_Nearby_Attractions_CityMap(Out Nearby_Attractions_CityMap)* operation formula.

Parameter	*Nearby_Attractions_CityMap*
Data Type	TABLE of Tourist_GPS_Coordinates: Text Map: Image Scenic_Spot: Text Scenic_Spot_GPS_Coordinates: Text End TABLE ;
Instances	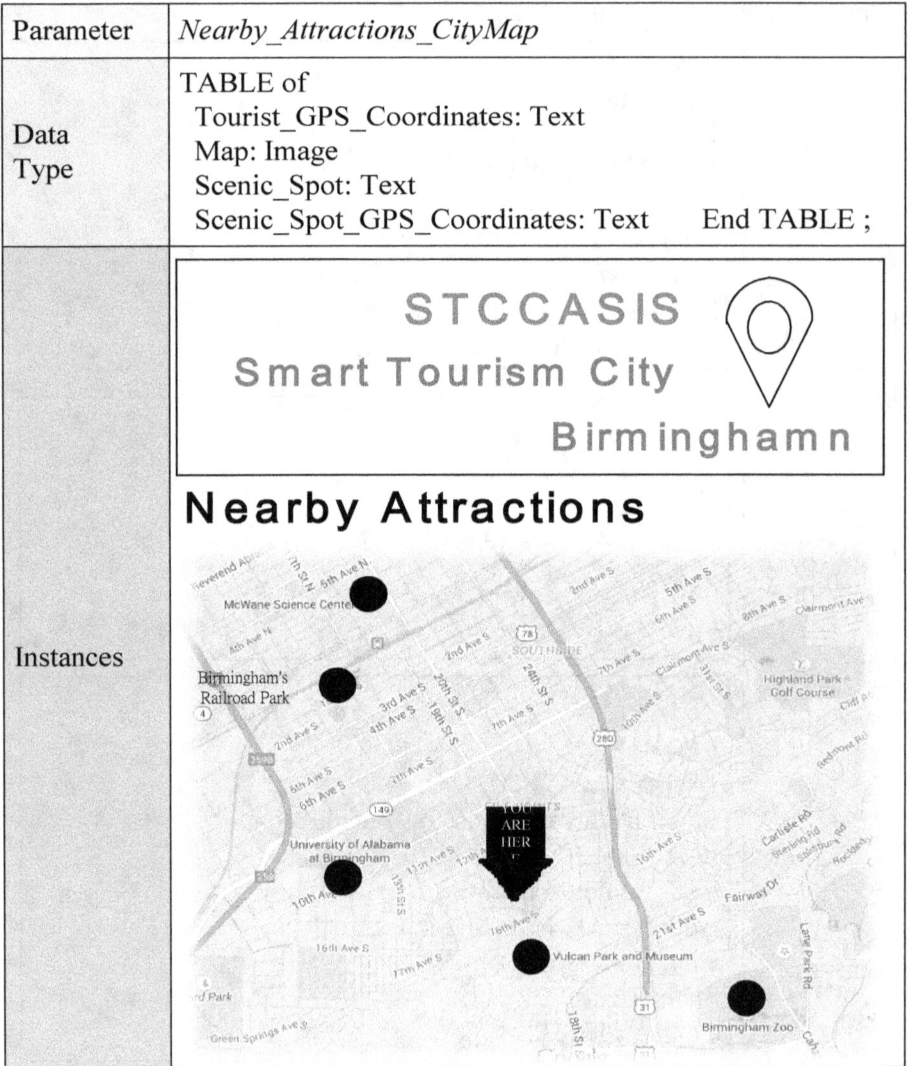

Figure 7-4 Composite Data Type Specification of *Nearby_Attractions_CityMap*

Figure 7-5 shows the primitive data type specification of the *Scenic_Spot*, *Tourist_GPS_Coordinates* parameters occurring in the *Show_Attraction_Details(In Scenic_Spot; Out Attraction_Details_Display)*, *SQL_Select_Nearby_Attractions(In Tourist_GPS_Coordinates; Out Nearby_Attractions_Query)*, *SQL_Select_Attraction_Details(In Scenic_Spot; Out Attraction_Details_Query)*, *Tourist_GPS_Positioning(Out Tourist_GPS_Coordinates)* and *Scenic_Spot_RFID_Positioning(Out Scenic_Spot)* operation formulas.

Parameter	Data Type	Instances
Scenic_ Spot	Text	Vulcan Park and Museum
Tourist_ GPS_ Coordinates	Text	33.490565,-86.794727

Figure 7-5 Primitive Data Type Specification

Figure 7-6 shows the composite data type specification of the *Attraction_Details_Display* output parameter occurring in the *Show_Attraction_Details(In Scenic_Spot; Out Attraction_Details_Display)* operation formula.

<table>
<tr><td>Parameter</td><td>Attraction_Details_Display</td></tr>
<tr><td>Data Type</td><td>TABLE of
Scenic_Spot: Text
Scenic_Spot_Address: Text
Description: Text
Main_Image: Image End TABLE ;</td></tr>
<tr><td>Instances</td><td>Vulcan Park and Museum
1701 Valley View Dr, Birmingham, AL 35209, U.S.A.
<table><tr><th>Description</th><th>Main_Image</th></tr><tr><td>Vulcan is the world's largest cast iron statue; made of 100,000 pounds of iron and 56 feet tall, he stands at the top of Red Mountain overlooking the city of Birmingham. But Vulcan is more than just a statue—Vulcan Park and Museum features spectacular views of Birmingham, an interactive history museum that examines Vulcan and Birmingham's story, a premier venue for private events, and a beautiful public park for visitors and locals to enjoy. With an official information center operated by the Greater Birmingham Convention and Visitors Bureau, Vulcan Park and Museum serves as the first stop for visitors to the Birmingham area!
-- Courtesy of visitvulcan.com --</td><td></td></tr></table></td></tr>
</table>

Figure 7-6 Composite Data Type Specification of *Attraction_Details_Display*

Figure 7-7 shows the composite data type specification of the *Personalized_Itinerary_Form* input parameter occurring in the *Input_Personalized_Itinerary(In Personalized_Itinerary_Form)* operation formula.

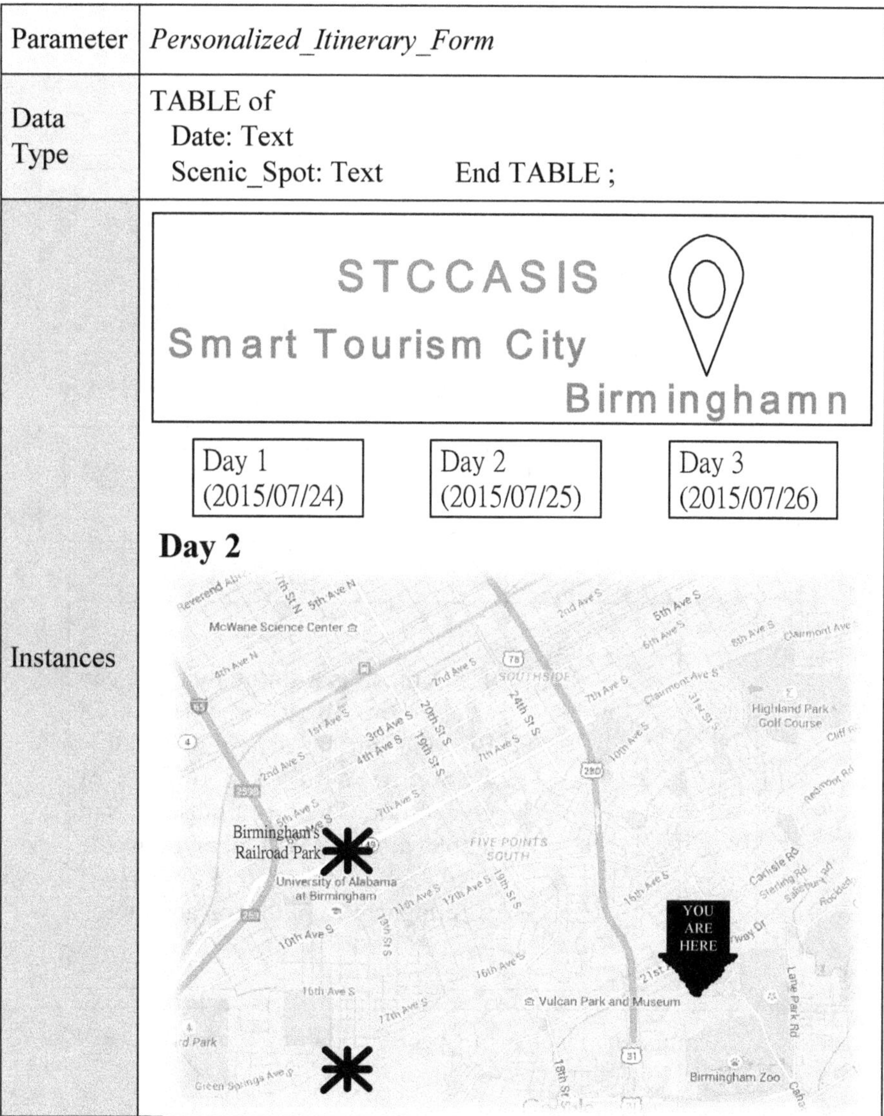

Parameter	*Personalized_Itinerary_Form*
Data Type	TABLE of Date: Text Scenic_Spot: Text End TABLE ;
Instances	STCCASIS Smart Tourism City Birminghamn Day 1 (2015/07/24) Day 2 (2015/07/25) Day 3 (2015/07/26) **Day 2**

Figure 7-7 Composite Data Type Specification of *Personalized_Itinerary_Form*

Figure 7-8 shows the composite data type specification of the *Recommending_Form* input parameter occurring in the *Scenic_Spot_Recommend(In Recommending_Form)* operation formula.

Parameter	*Recommending_Form*
Data Type	TABLE of Stars: Integer Comments: Text End TABLE ;
Instances	STCCASIS Smart Tourism City Birminghamn **Checking In And Recommending** Vulcan Park and Museum Stars: 1, 2, 3, 4 (selected), 5 Comments: The Vulcan is a statue that you can see from about anyplace in Birmingham. It's on a high hill. It is a $6 entry with a kid and senior citizen rate. There is a museum and an elevator that will take you to the top of the statue. See all of Birmingham from the top and there.

Figure 7-8 Composite Data Type Specification of *Recommending_Form*

Figure 7-9 shows the composite data type specification of the *New_Account_Query* input parameter occurring in the *SQL_Insert_New_Account(In New_Account_Query)* operation formula.

Parameter	*New_Account_Query*
Data Type	TABLE of Username: Text Email_Address: Text First_Name: Text Last_Name: Text Address: Text City: Text State: Text Country: Text End TABLE ;
Instances	(see table below)

Username	Email_ Address	First_ Name	Last_ Name	Address	City	State	Country
A1B2C3D4	adolph6789 @gmail.com	Adolph	Bryant	702 Ross Street	Dallas	TX	U.S.A.

Figure 7-9 Composite Data Type Specification of *New_Account_Query*

Figure 7-10 shows the composite data type specification of the *Nearby_Attractions_Query* output parameter occurring in the *SQL_Select_Nearby_Attractions(In Tourist_GPS_Coordinates; Out Nearby_Attractions_Query)* operation formula.

<table>
<tr><td>Parameter</td><td>Nearby_Attractions_Query</td></tr>
<tr><td>Data Type</td><td>TABLE of
Scenic_Spot: Text
Scenic_Spot_GPS_Coordinates: Text
End TABLE ;</td></tr>
<tr><td>Instances</td><td>
<table>
<tr><th>Scenic_Spot</th><th>Scenic_Spot_ GPS_Coordinates</th></tr>
<tr><td>Birmingham Zoo</td><td>33.48657862, -86.77911758</td></tr>
<tr><td>Vulcan Park and Museum</td><td>33.490565, -86.794727</td></tr>
<tr><td>McWane Science Center</td><td>33.51520752, -86.80830002</td></tr>
<tr><td>Birmingham's Railroad Park</td><td>33.5099169, -86.8084382</td></tr>
<tr><td>University of Alabama at Birmingham</td><td>33.49302095, -86.80898666</td></tr>
</table>
</td></tr>
</table>

Figure 7-10 Composite Data Type Specification of *Nearby_Attractions_Query*

Figure 7-11 shows the composite data type specification of the *Attraction_Details_Query* output parameter occurring in the *SQL_Select_Attraction_Details(In Scenic_Spot; Out Attraction_Details_Query)* operation formula.

<table>
<tr><td>Parameter</td><td>Attraction_Details_Query</td></tr>
<tr><td>Data Type</td><td>TABLE of
Scenic_Spot: Text
Scenic_Spot_Address: Text
Description: Text
Main_Image: ImageData
End TABLE ;</td></tr>
<tr><td>Instances</td><td>
<table>
<tr><th>Scenic_Spot</th></tr>
<tr><td>Vulcan Park and Museum</td></tr>
</table>
<table>
<tr><th>Scenic_Spot_Address</th></tr>
<tr><td>1701 Valley View Dr,
Birmingham, AL 35209,
U.S.A.</td></tr>
</table>
<table>
<tr><th>Description</th><th>Main_Image</th></tr>
<tr><td>Vulcan is the world's largest cast iron statue; made of 100,000 pounds of iron and 56 feet tall, he stands at the top of Red Mountain overlooking the city of Birmingham. But Vulcan is more than just a statue—Vulcan Park and Museum features spectacular views of Birmingham, an interactive history museum that examines Vulcan and Birmingham's story, a premier venue for private events, and a beautiful public park for visitors and locals to enjoy. With an official information center operated by the Greater Birmingham Convention and Visitors Bureau, Vulcan Park and Museum serves as the first stop for visitors to the Birmingham area!
-- Courtesy of visitvulcan.com --</td><td>011101000010101010
010101001000000010
010010101001010010
101010010101001010
010010100101110100
001010101001010100
100000001001001010
100100101010010100
101010100101010010
100100101001011101
000010101010010101
001000000010010010
101001010010101010
010101001010010010
100101110100001010
101001010100100000</td></tr>
</table>
</td></tr>
</table>

Figure 7-11 Composite Data Type Specification of *Attraction_Details_Query*

Figure 7-12 shows the composite data type specification of the *Personalized_Itinerary_Query* input parameter occurring in the *SQL_Insert_Personalized_Itinerary(In Personalized_Itinerary_Query)* operation formula.

<table>
<tr><td>Parameter</td><td>Personalized_Itinerary_Query</td></tr>
<tr><td>Data Type</td><td>TABLE of
  Date: Text
  Scenic_Spot: Text
End TABLE ;</td></tr>
<tr><td>Instances</td><td>
<table>
<tr><th>Date</th><th>Scenic_Spot</th></tr>
<tr><td>20150724</td><td>Birmingham Zoo</td></tr>
<tr><td>20150724</td><td>Vulcan Park and Museum</td></tr>
<tr><td>20150725</td><td>McWane Science Center</td></tr>
<tr><td>20150725</td><td>Birmingham's Railroad Park</td></tr>
<tr><td>20150726</td><td>University of Alabama at Birmingham</td></tr>
</table>
</td></tr>
</table>

Figure 7-12 Composite Data Type Specification of *Personalized_Itinerary_Query*

Figure 7-13 shows the composite data type specification of the *Checking_In_And_Recommending_Query* input parameter occurring in the *SQL_Insert_Checking_In_And_Recommending(In Checking_In_And_Recommending_Query)* operation formula.

<table>
<tr><td>Parameter</td><td>Checking_In_And_Recommending_Query</td></tr>
<tr><td>Data Type</td><td>TABLE of
Scenic_Spot: Text
First_Name: Text
Last_Name: Text
Stars: Integer
Comments: Text End TABLE ;</td></tr>
<tr><td>Instances</td><td>
<table>
<tr><th>Scenic_Spot</th></tr>
<tr><td>Vulcan Park and Museum</td></tr>
</table>
<table>
<tr><th>First Name</th><th>Last Name</th><th>Stars</th><th>Comments</th></tr>
<tr><td>Edgar</td><td>Ashworth</td><td>4</td><td>The Vulcan is a statue that you can see from about anyplace in Birmingham. It's on a high hill. It is a $6 entry with a kid and senior citizen rate. There is a museum and an elevator that will take you to the top of the statue. See all of Birmingham from the top and there.</td></tr>
</table>
</td></tr>
</table>

Figure 7-13 Composite Data Type Specification of *Checking_In_And_Recommending_Query*

7-3 Interaction Flow Diagram

The overall behavior of the *Smart Tourism City Cloud Applications and Services IoT System* (STCCASIS) includes the *Creating_New_Account*, *Showing_Nearby_Attractions_CityMap*, *Extracting_Attraction_Details*, *Planning_Personalized_Itinerary* and *Scenic_Spot_Checking_In_And_Recommending* behaviors. Each individual behavior is represented by an execution path. Systems design 2.0 uses an IFD to design each one of these execution paths.

Figure 7-14 shows an IFD of the *Creating_New_Account* behavior. First, actor *Tourist* interacts with the *New_Account_UI* component through the *Input_New_Account* operation call interaction, carrying the *New_Account_Form* input parameter. Next, component *New_Account_GUI* interacts with the *STCCASIS_Database* component through the *SQL_Insert_New_Account* operation call interaction, carrying the *New_Account_Query* input parameter.

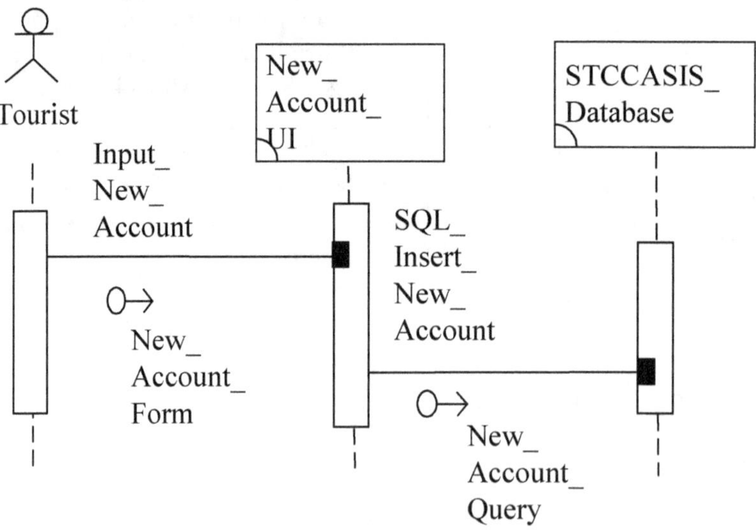

Figure 7-14 IFD of the *Creating_New_Account* Behavior

Figure 7-15 shows an IFD of the *Showing_Nearby_Attractions_CityMap* behavior. First, actor *Tourist* interacts with the *Nearby_Attractions_CityMap_UI* component through the *Show_Nearby_Attractions_CityMap* operation call interaction. Next, component *Nearby_Attractions_CityMap_UI* interacts with the *Tourist_GPS_M (M = AAA0000 to ZZZ9999)* component through the *Tourist_GPS_Positioning* operation call interaction, carrying the *Tourist_GPS_Coordinates* output parameter. Continuingly, component *Nearby_Attractions_CityMap_UI* interacts with the *STCCASIS_Database* component through the *SQL_Select_Nearby_Attractions* operation call interaction, carrying the *Tourist_GPS_Coordinates* input parameter and *Nearby_Attractions_Query* output parameter. Finally, actor *Tourist* interacts with the *Nearby_Attractions_CityMap_UI* component through the *Show_Nearby_Attractions_CityMap* operation return interaction, carrying the *Nearby_Attractions_CityMap* output parameter.

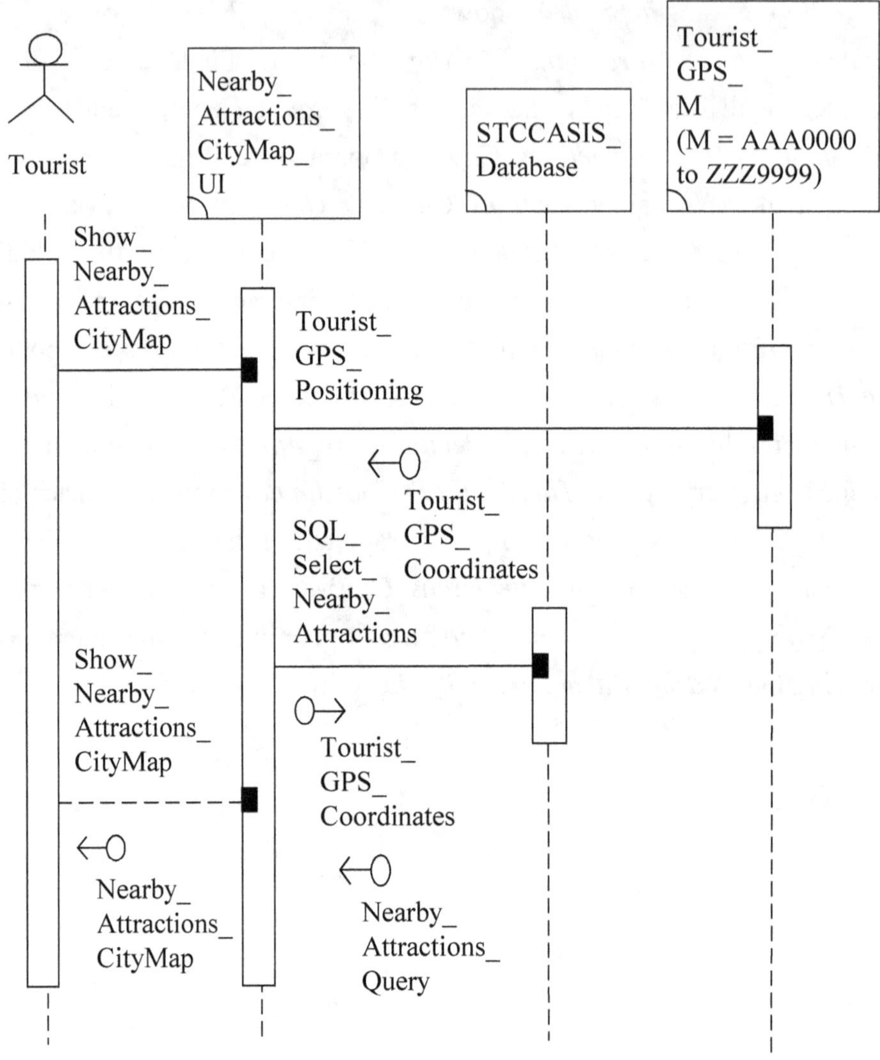

Figure 7-15 IFD of the *Showing_Nearby_Attractions_CityMap* Behavior

Figure 7-16 shows an IFD of the *Extracting_Attraction_Details* behavior. First, actor *Tourist* interacts with the *Attraction_Details_UI* component through the *Show_Attraction_Details* operation call interaction, carrying the *Scenic_Spot* input parameter. Next, component *Attraction_Details_UI* interacts with the *STCCASIS_Database* component through the *SQL_Select_Attraction_Details* operation call interaction, carrying the *Scenic_Spot* input parameter and *Attraction_Details_Query* output parameter. Finally, actor *Tourist* interacts with the *Attraction_Details_UI* component through the *Show_Attraction_Details* operation return interaction, carrying the *Attraction_Details_Display* output parameter.

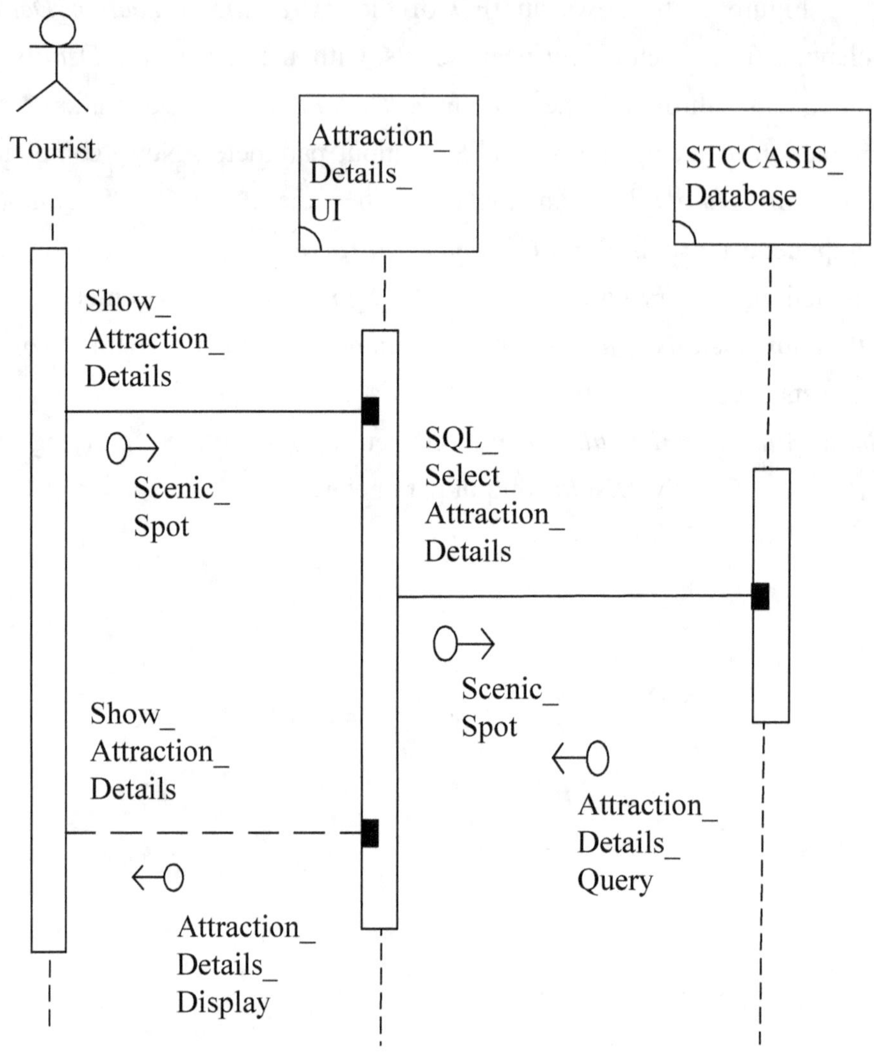

Figure 7-16 IFD of the *Extracting_Attraction_Details* Behavior

Figure 7-17 shows an IFD of the *Planning_Personalized_Itinerary* behavior. First, actor *Tourist* interacts with the *Personalized_Itinerary_UI* component through the *Input_Personalized_Itinerary* operation call interaction, carrying the

Personalized_Itinerary_Form input parameter. Finally, component *Personalized_Itinerary_UI* interacts with the *STCCASIS_Database* component through the *SQL_Insert_Personalized_Itinerary* operation call interaction, carrying the *Personalized_Itinerary_Query* input parameter.

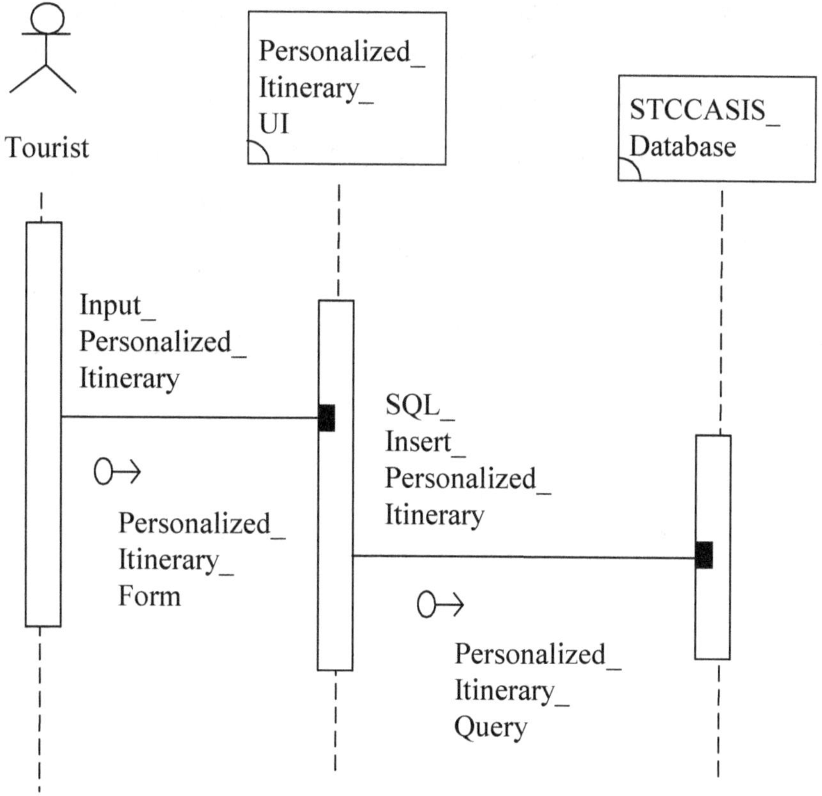

Figure 7-17 IFD of the *Planning_Personalized_Itinerary* Behavior

Figure 7-18 shows an IFD of the *Scenic_Spot_Checking_In_And_Recommending* behavior. First, actor *Tourist* interacts with the *Checking_In_And_Recommending_UI* component through the *Scenic_Spot_Check_In* operation call interaction. Next, component *Checking_In_And_Recommending_UI* interacts with the *Scenic_Spot_RFID_N (N = 000 to 999)* component through the *Scenic_Spot_RFID_Positioning* operation call interaction, carrying the *Scenic_Spot* output parameter. Continuingly, actor *Tourist* interacts with the *Checking_In_And_Recommending_UI* component through the *Scenic_Spot_Recommend* operation call interaction, carrying the *Recommending_Form* input parameter. Finally, component *Checking_In_And_Recommending_UI* interacts with the *STCCASIS_Database* component through the *SQL_Insert_Checking_In_And_Recommending* operation call interaction, carrying the *Checking_In_And_Recommending_Query* input parameter.

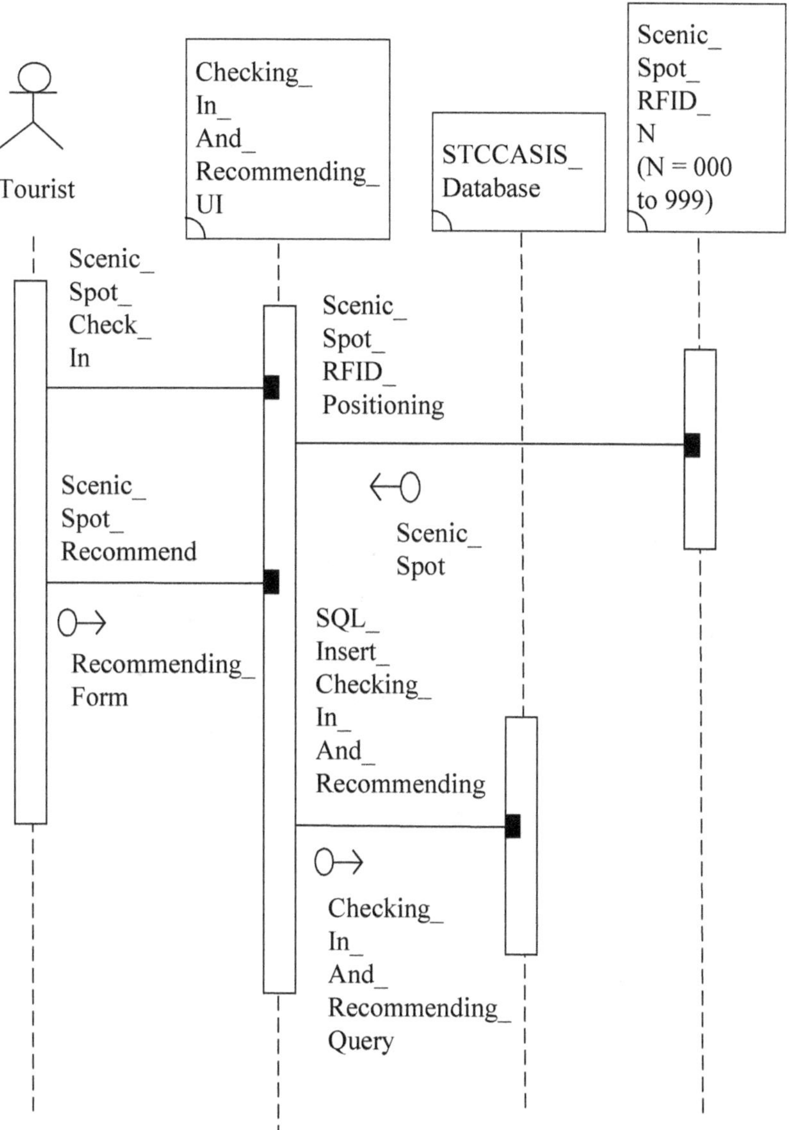

Figure 7-18 IFD of the *Scenic_Spot_Checking_In_And_Recommending* Behavior

Chapter 8: Systems Design 2.0 of the Home Care Cloud Applications and Services IoT System

Home care, also referred to as domiciliary care, is supportive care provided in the home [Shyu02]. Care may be provided by licensed healthcare professionals who provide medical care needs or by professional caregivers who provide daily care to help to ensure the activities of daily living are met.

Advancements in cloud applications and services IoT systems present enormous potential for the intensified healthcare support of senior residents at home. By employing these systems in the home, senior residents are able to live independently for a longer period of time, helping to reduce costs and the need for additional caregiver resources in the process.

A home care cloud applications and services IoT system (HCCASIS) that would work well for home monitoring should include position sensors which will sense the positioning data of senior residents at home, supposedly in every 15 to 20 seconds. The advantage of those positioning data is that they are highly precise and deliver 3-dimensional data, accurate to 2.5 centimeters. HCCASIS will be constantly monitoring these measurements which would show the movements and locations of each senior resident at home. An alert will be triggered if: a) a resident falls on the floor for more than a certain amount of time, or b) a resident has irregular movement pattern, or c) a resident shows a significant lack of movement. As soon as any alert is triggered, HCCASIS will send out an alert notification to relatives, a home care provider, or security call center for help.

Behaviors of HCCASIS consist of: a) behavior of *Registering_Home_Account*, b) behavior of *Sensing_Residents_Position*, c) behavior of *Alerts_Notifying*, d) behavior of

Recording_Emergency_Responses and e) behavior of *Printing_Monthly_Statistics*.

Using the structure-behavior coalescence (SBC) approach, we shall go through: a) framework diagram, b) component operation diagram and c) interaction flow diagram, to accomplish the systems design 2.0 for HCCASIS.

8-1 Framework Diagram

Systems design 2.0 uses a framework diagram (FD) to design the multi-layer composition and decomposition of the *Home Care Cloud Applications and Services IoT System* (HCCASIS) as shown in Figure 8-1.

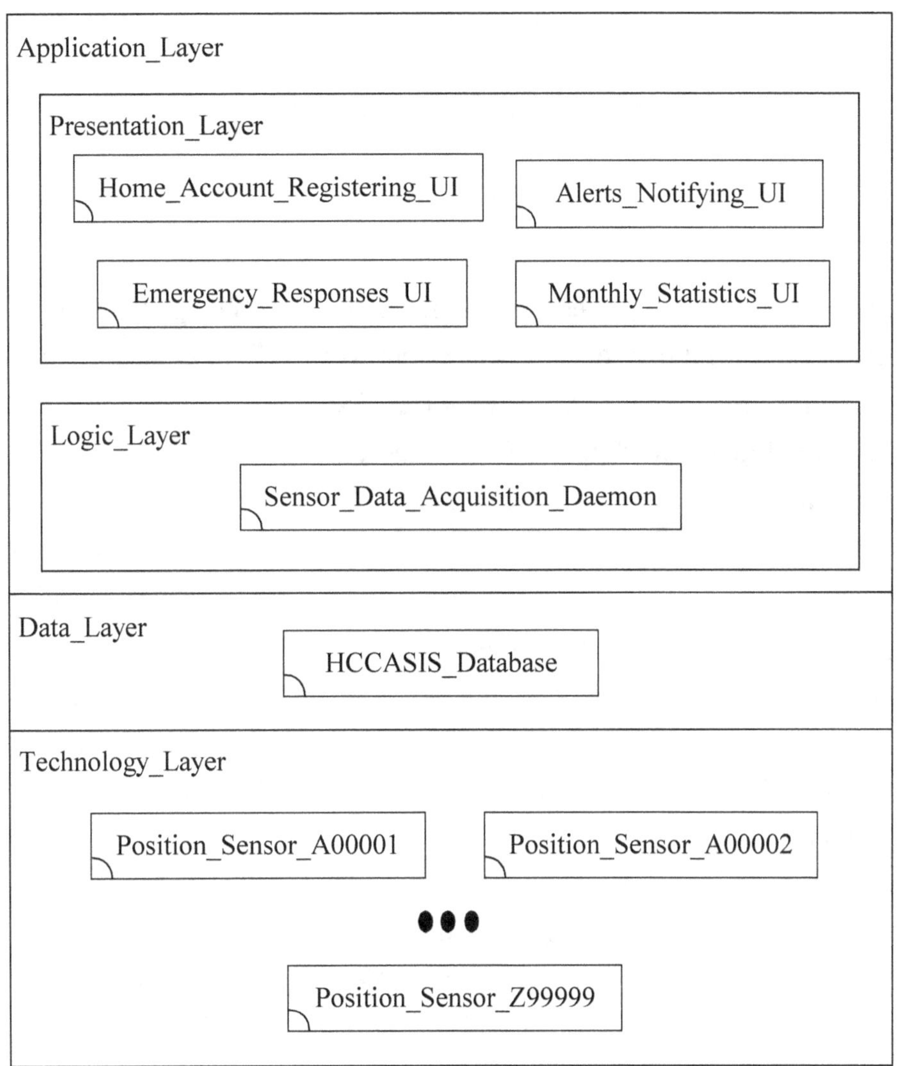

Figure 8-1 FD of the *HCCASIS*

In the above figure, *Presentation_Layer* and *Logic_Layer* are sub-layers of *Application_Layer*. *Presentation_Layer* contains the *Home_Account_Registering_UI*, *Alerts_Notifying_UI*, *Emergency_Responses_UI* and *Monthly_Statistics_UI* components; *Logic_Layer* contains the *Sensor_Data_Acquisition_Daemon* component;

Data_Layer contains the *HCCASIS_Database* component; *Technology_Layer* contains the *Position_Sensor_A00001*, *Position_Sensor_A00002,…, Position_Sensor_Z99999* components.

8-2 Component Operation Diagram

Systems design 2.0 uses a component operation diagram (COD) to design the operations of all components of the *Home Care Cloud Applications and Services IoT System* (HCCASIS) as shown in Figure 8-2. In the figure, component *Home_Account_Registering_UI* has one operation: *Input_Home_Data*; component *Alerts_Notifying_UI* has two operations: *Showing_All_Alerts* and *Displaying_Alerts*; component *Emergency_Responses_UI* has one operation: *Input_Emergency_Responses*; component *Monthly_Statistics_UI* has one operation: *PrintButton_Click*; component *Sensor_Data_Acquisition_Daemon* has one operation: *Fork_SDAD_Process*; component *HCCASIS_Database* has five operations: *SQL_Insert_Home_Data*, *SQL_Insert_3-Dimensional_Locations*, *SQL_Select_3-Dimensional_Locations_for_Alerts_Analysis*, *SQL_Insert_Emergency_Responses* and *SQL_Select_Monthly_Statistics*; component *Position_Sensor_N* (N = A00001 to Z99999) has two operations: *Sensing_Position* and *Returning_Position.*

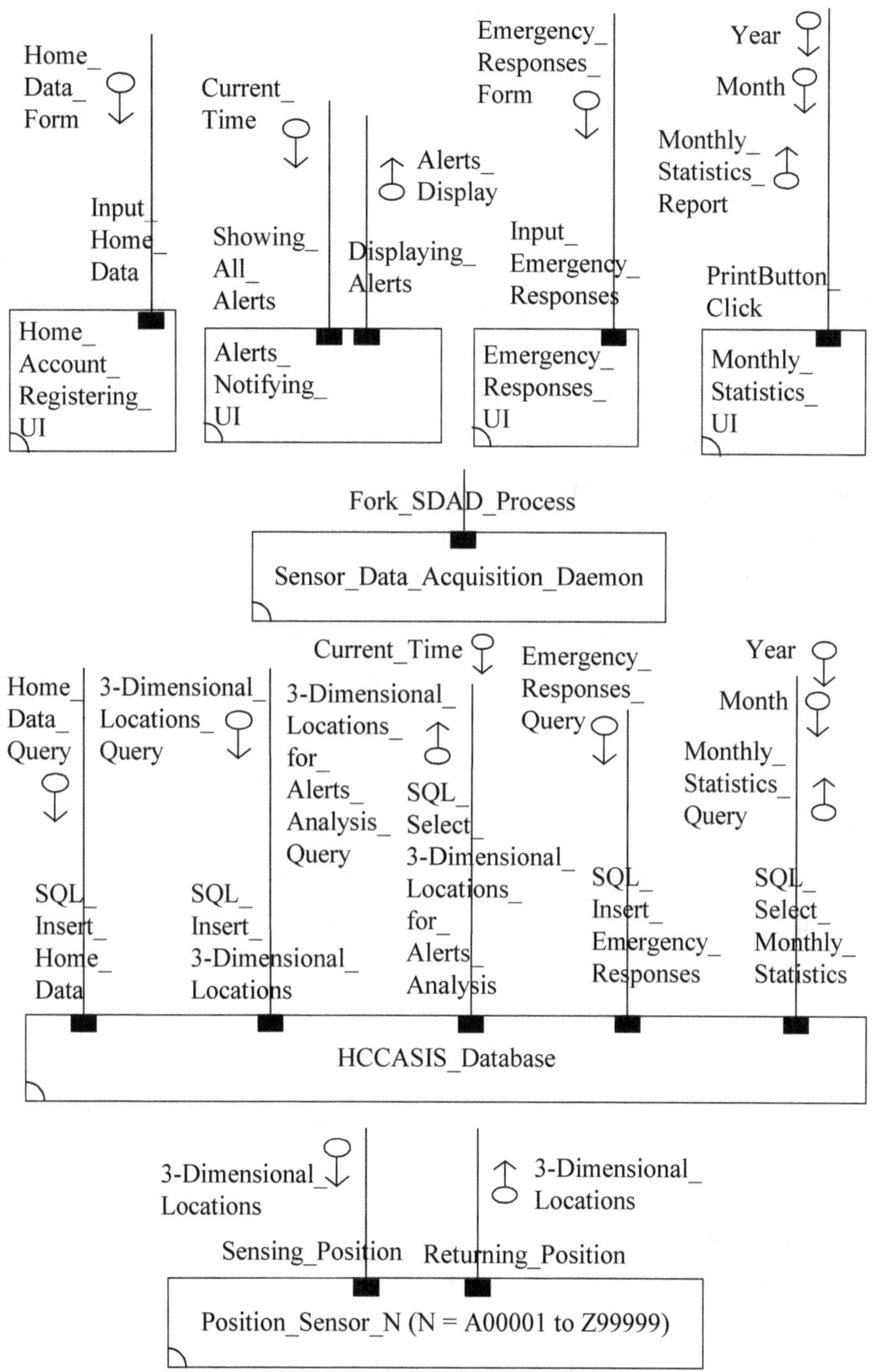

Figure 8-2 COD of the *HCCASIS*

The operation formula of *Input_Home_Data* is *Input_Home_Data(In Home_Data_Form)*. The operation formula of *Showing_All_Alerts* is *Showing_All_Alerts(In Current_Time)*. The operation formula of *Displaying_Alerts* is *Displaying_Alerts(Out Alerts_Display)*. The operation formula of *Input_Emergency_Responses* is *Input_Emergency_Responses(In Emergency_Responses_Form)*. The operation formula of *PrintButton_Click* is *PrintButton_Click(In Year, Month*; *Out Monthly_Statistics_Report)*. The operation formula of *Fork_SDAD_Process* is *Fork_SDAD_Process*. The operation formula of *SQL_Insert_Home_Data* is *SQL_Insert_Home_Data(In Home_Data_Query)*. The operation formula of *SQL_Insert_3-Dimensional_Locations* is *SQL_Insert_3-Dimensional_Locations(In 3-Dimensional_Locations_Query)*. The operation formula of *SQL_Select_3-Dimensional_Locations_for_Alerts_Analysis* is *SQL_Select_3-Dimensional_Locations_for_Alerts_Analysis(In Current_Time*; *Out 3-Dimensional_Locations_for_Alerts_Analysis_Query)*. The operation formula of *SQL_Insert_Emergency_Responses* is *SQL_Insert_Emergency_Responses(In Emergency_Responses_Query)*. The operation formula of *SQL_Select_Monthly_Statistics* is *SQL_Select_Monthly_Statistics(In Year, Month; Out Monthly_Statistics_Query)*. The operation formula of *Sensing_Position* is *Sensing_Position(In 3-Dimensional_Locations)*. The operation formula of *Returning_Position* is *Returning_Position(Out 3-Dimensional_Locations)*.

Figure 8-3 shows the composite data type specification of the *Home_Data_Form* input parameter occurring in the *Input_Home_Data(In Home_Data_Form)* operation formula.

Parameter	*Home_Data_Form*
Data Type	TABLE of Home_No: Text Address: Text Relative Name: Text Relative Phone: Text First_Name: Text Last_Name: Text Age: Integer End TABLE ;
Instances	HCCASIS HomeCare Services **Home Data Form** Home_No: A00001 Address: 8417 Lorna Rd, Birmingham, AL 35216 Relative Name: Tom Hutchison Relative Phone : (205) 786-4328 First_Name Last_Name Age Grace Hutchison 82 John Hutchison 83

Figure 8-3 Composite Data Type Specification of *Home_Data_Form*

Figure 8-4 shows the primitive data type specification of the *Current_Time*, *Year*, *Month* parameters occurring in the *Showing_All_Alerts(In Current_Time)*, *PrintButton_Click(In Year, Month; Out Monthly_Statistics_Report)*, *SQL_Select_Alerts(In Current_Time; Out Alerts_Query)* and *SQL_Select_Monthly_Statistics(In Year, Month; Out Monthly_Statistics_Query)* operation formulas.

Parameter	Data Type	Instances
Current_ Time	Text	20150612231745
Year	Text	2015
Month	Text	06

Figure 8-4   Primitive Data Type Specification

Figure 8-5 shows the composite data type specification of the *Alerts_Display* output parameter occurring in the *Displaying_Alerts(Out Alerts_Display)* operation formula.

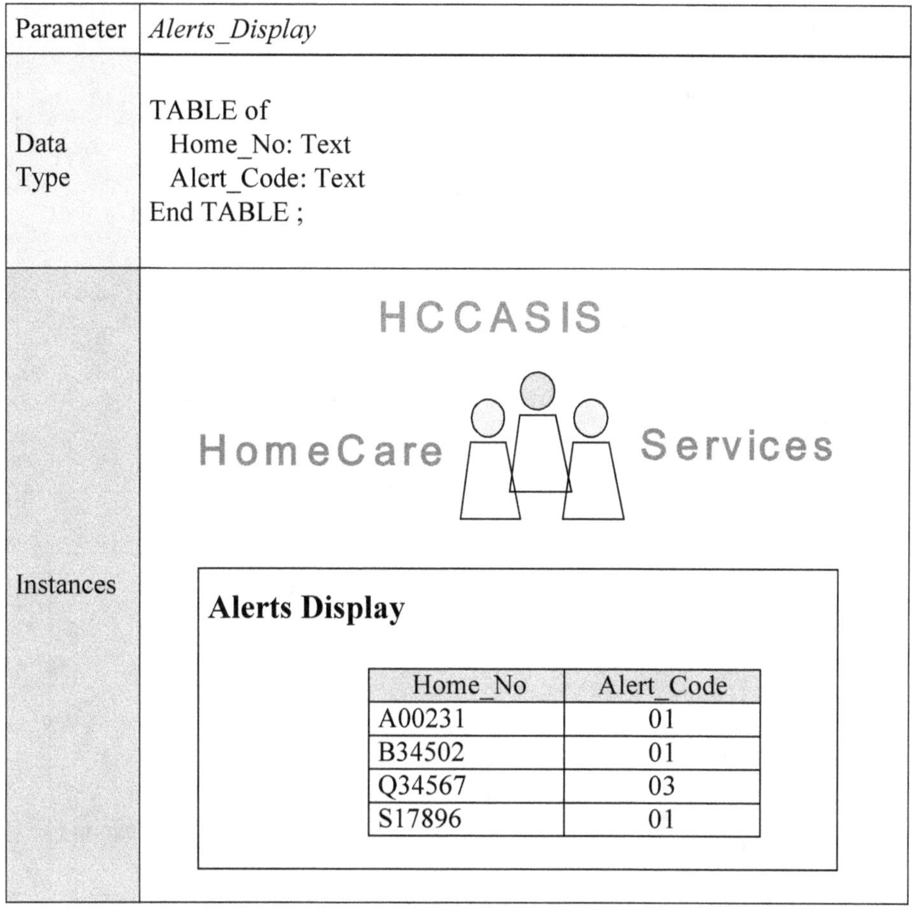

Parameter	*Alerts_Display*

Data Type	TABLE of
	  Home_No: Text
	  Alert_Code: Text
	End TABLE ;

Instances	HCCASIS HomeCare Services **Alerts Display**

Home_No	Alert_Code
A00231	01
B34502	01
Q34567	03
S17896	01

Figure 8-5 Composite Data Type Specification of *Alerts_Display*

Figure 8-6 shows the composite data type specification of the *Emergency_Responses_Form* input parameter occurring in the *Input_Emergency_Responses(In Emergency_Responses_Form)* operation formula.

Parameter	*Emergency_Responses_Form*

Data Type	TABLE of
	  Home_No: Text
	  Time_to_Respond: Text
	  Actions_Taken_to_Respond: Text
	End TABLE ;

Instances	HCCASIS HomeCare Services **Emergency Responses Form** Home_No: A12345 Time_to_Respond: 20150607134020 Actions_Taken_to_Respond * Send people there * Nofify the relatives

Figure 8-6 Composite Data Type Specification of *Emergency_Responses_Form*

Figure 8-7 shows the composite data type specification of the *Monthly_Statistics_Report* output parameter occurring in the *PrintButton_Click(In Year, Month; Out Monthly_Statistics_Report)* operation formula.

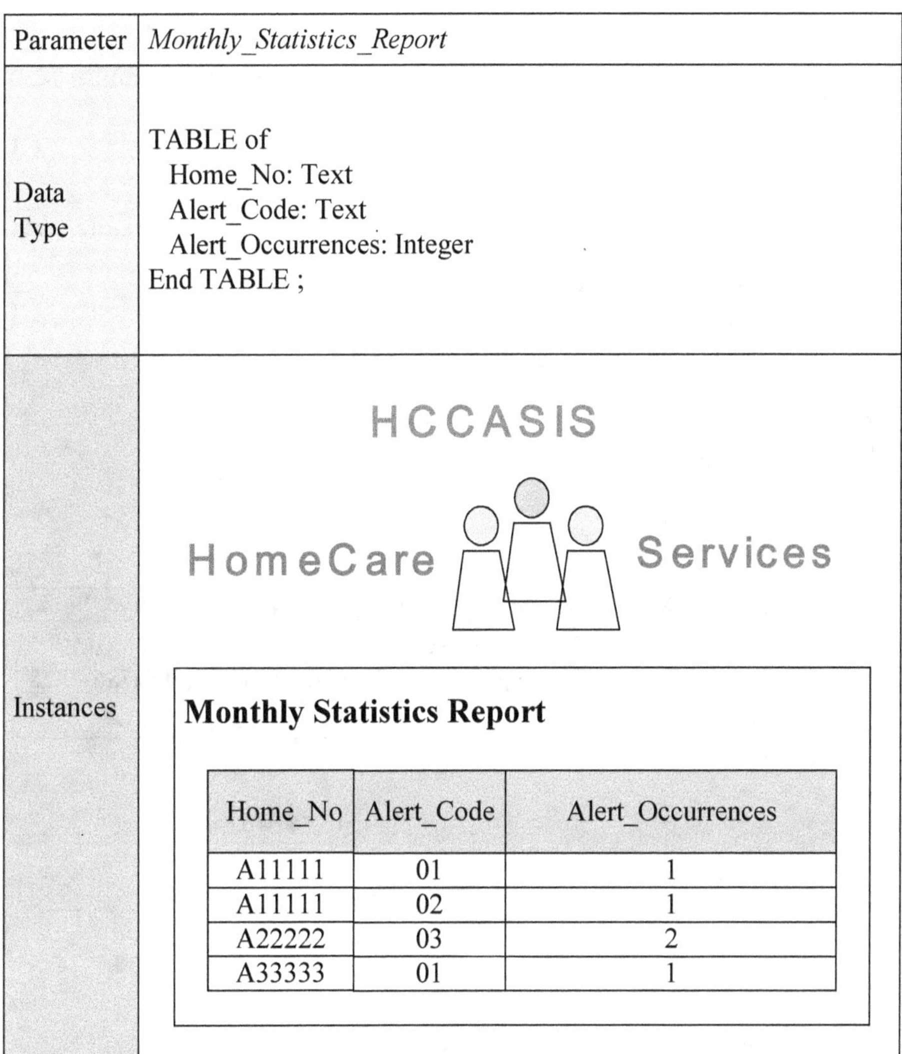

<table>
<tr><td>Parameter</td><td>Monthly_Statistics_Report</td></tr>
<tr><td>Data Type</td><td>TABLE of
  Home_No: Text
  Alert_Code: Text
  Alert_Occurrences: Integer
End TABLE ;</td></tr>
<tr><td>Instances</td><td>HCCASIS
HomeCare Services

Monthly Statistics Report
<table>
<tr><th>Home_No</th><th>Alert_Code</th><th>Alert_Occurrences</th></tr>
<tr><td>A11111</td><td>01</td><td>1</td></tr>
<tr><td>A11111</td><td>02</td><td>1</td></tr>
<tr><td>A22222</td><td>03</td><td>2</td></tr>
<tr><td>A33333</td><td>01</td><td>1</td></tr>
</table></td></tr>
</table>

Figure 8-7 Composite Data Type Specification of *Monthly_Statistics_Report*

Figure 8-8 shows the composite data type specification of the *Home_Data_Query* output parameter occurring in the *SQL_Insert_Home_Data(In Home_Data_Query)* operation formula.

<table>
<tr><td>Parameter</td><td>Home_Data_Query</td></tr>
<tr><td>Data Type</td><td>TABLE of
Home_No: Text
Address: Text
Relative Name: Text
Relative Phone: Text
First_Name: Text
Last_Name: Text
Age: Integer
End TABLE ;</td></tr>
<tr><td>Instances</td><td>
<table>
<tr><th>Home_No</th><th>Address</th><th>Relative Name</th><th>Relative Phone</th></tr>
<tr><td>A00001</td><td>8417 Lorna Rd,
Birmingham,
AL 35216</td><td>Tom Hutchison</td><td>(205)786-4328</td></tr>
</table>
<table>
<tr><th>First_Name</th><th>Last_Name</th><th>Age</th></tr>
<tr><td>Grace</td><td>Hutchison</td><td>82</td></tr>
<tr><td>John</td><td>Hutchison</td><td>83</td></tr>
</table>
</td></tr>
</table>

Figure 8-8 Composite Data Type Specification of *Home_Data_Query*

Figure 7-8 shows the composite data type specification of the *3-Dimensional_Locations_Query* input parameter occurring in the *SQL_Insert_3-Dimensional_Locations(In 3-Dimensional_Locations_Query)* operation formula.

Parameter	*3-Dimensional_Locations_Query*
Data Type	TABLE of Home_No: Text Recorded_Time: Text X-coordinate: Real Y-coordinate: Real Z-coordinate: Real End TABLE ;
Instances	see tables below

Home_No	Recorded_Time
A12345	20150606142530

X-coordinate	Y-coordinate	Z-coordinate
240	120	38
200	150	31

Figure 8-9 Composite Data Type Specification of *3-Dimensional_Locations_Query*

Figure 8-10 shows the composite data type specification of the *3-Dimensional_Locations_for_Alerts_Analysis_Query* output parameter occurring in the *SQL_Select_3-Dimensional_Locations_for_Alerts_Analysis(In Current_Time; Out 3-Dimensional_Locations_for_Alerts_Analysis_Query)* operation formula.

<table>
<tr><td>Parameter</td><td>3-Dimensional_Locations_for_Alerts_Analysis_Query</td></tr>
<tr><td>Data Type</td><td>TABLE of
Home_No: Text
Recorded_Time: Text
X-coordinate: Real
Y-coordinate: Real
Z-coordinate: Real
End TABLE ;</td></tr>
<tr><td>Instances</td><td>
<table>
<tr><td>Home_ No</td><td>Recorded_ Time</td><td>X- coordinate</td><td>Y- coordinate</td><td>Z- coordinate</td></tr>
<tr><td>A11111</td><td>20150606142500</td><td>200</td><td>150</td><td>38</td></tr>
<tr><td>A11111</td><td>20150606142530</td><td>180</td><td>140</td><td>38</td></tr>
<tr><td>A22222</td><td>20150606142500</td><td>100</td><td>250</td><td>30</td></tr>
<tr><td>A22222</td><td>20150606142530</td><td>100</td><td>240</td><td>30</td></tr>
</table>
</td></tr>
</table>

Figure 8-10 Composite Data Type Specification of *3-Dimensional_Locations_for_Alerts_Analysis_Query*

Figure 8-11 shows the composite data type specification of the *Emergency_Responses_Query* input parameter occurring in the *SQL_Insert_Emergency_Responses(In Emergency_Responses_Query)* operation formula.

Parameter	*Emergency_Responses_Query*
Data Type	TABLE of Home_No: Text Time_to_Respond: Text Actions_Taken_to_Respond: Text End TABLE ;
Instances	| Home_No | Time_to_Respond | | A12345 | 20150607134020 | | Actions_Taken_to_Respond | | Send people there | | Nofify the relatives |

Figure 8-11 Composite Data Type Specification of *Emergency_Responses_Query*

Figure 8-12 shows the composite data type specification of the *Monthly_Statistics_Query* output parameter occurring in the *SQL_Select_Monthly_Statistics(In Year, Month; Out Monthly_Statistics_Query)* operation formula.

<table>
<tr><td>Parameter</td><td>Monthly_Statistics_Query</td></tr>
<tr><td>Data Type</td><td>TABLE of
  Home_No: Text
  Alert_Occurrence_Time: Text
  Alert_Code: Text
End TABLE ;</td></tr>
<tr><td>Instances</td><td>
<table>
<tr><th>Home_No</th><th>Alert_Occurrence_Time</th><th>Alert_Code</th></tr>
<tr><td>A11111</td><td>20150603142500</td><td>01</td></tr>
<tr><td>A11111</td><td>20150606091200</td><td>02</td></tr>
<tr><td>A22222</td><td>20150602183030</td><td>03</td></tr>
<tr><td>A33333</td><td>20150606142500</td><td>01</td></tr>
</table>
</td></tr>
</table>

Figure 8-12 Composite Data Type Specification of *Monthly_Statistics_Query*

Figure 8-13 shows the composite data type specification of the *3-Dimensional_Locations* parameters occurring in the *Sensing_Position(In 3-Dimensional_Locations)* and *Returning_Position(Out 3-Dimensional_Locations)* operation formulas.

Parameter	*3-Dimensional_Locations*
Data Type	TABLE of X-coordinate: Real Y-coordinate: Real Z-coordinate: Real End TABLE ;
Instances	(nested table below)

X-coordinate	Y-coordinate	Z-coordinate
240	120	37
200	150	30

Figure 8-13 Composite Data Type Specification of *3-Dimensional_Locations*

8-3 Interaction Flow Diagram

The overall behavior of the *Home Care Cloud Applications and Services IoT System* (HCCASIS) includes five individual behaviors: *Registering_Home_Account*, *Sensing_Residents_Position*, *Alerts_Notifying*, *Recording_Emergency_Responses*, *Printing_Monthly_Statistics*. Each individual behavior is represented by an execution path. Systems design 2.0 uses an IFD to design each one of these execution paths.

Figure 8-14 shows an IFD of the *Registering_Home_Account* behavior. First, actor *Homecare_Provider* interacts with the *Home_Account_Registering_UI* component through the

Input_Home_Data operation call interaction, carrying the *Home_Data_Form* input parameter. Next, component *Home_Account_Registering_UI* interacts with the *HCCASIS_Database* component through the *SQL_Insert_Home_Data* operation call interaction, carrying the *Home_Data_Query* input parameter.

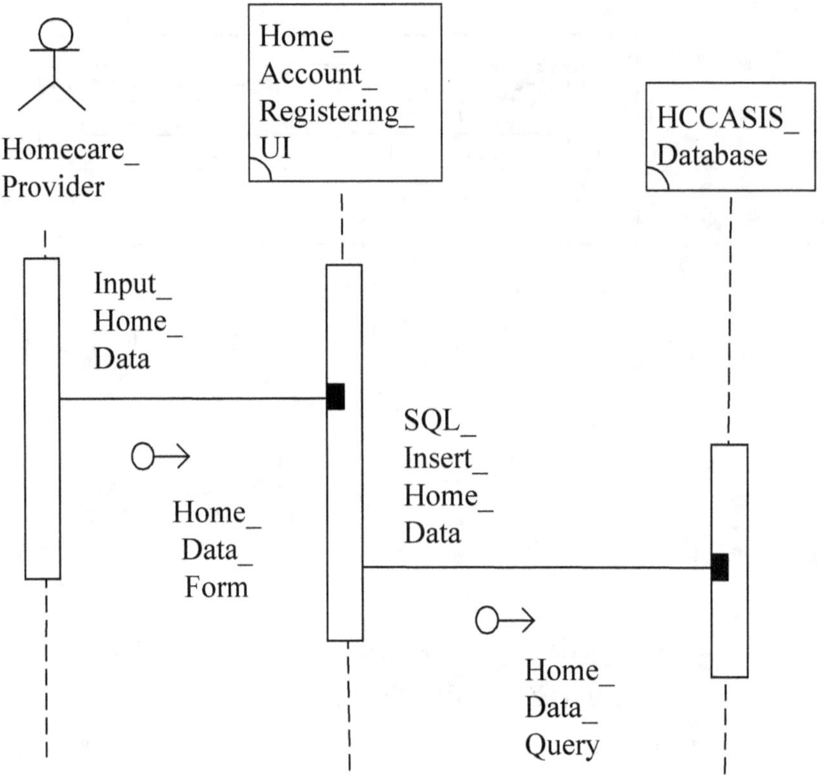

Figure 8-14 IFD of the *Registering_Home_Account* Behavior

Figure 8-15 shows an IFD of the *Sensing_Residents_Position* behavior. First, actor *Server_Root* interacts with the *Sensor_Data_Acquisition_Daemon* component through the *Fork_SDAD_Process* operation call interaction. Next, actor

Senior_Residents interacts with the *Position_Sensor_N* (N = A00001 to Z99999) component through the *Sensing_Position* operation call interaction, carrying the *3-Dimensional_Locations* input parameter. Continuingly, component *Sensor_Data_Acquisition_Daemon* interacts with the *Position_Sensor_N* (N = A00001 to Z99999) component through the *Returning_Position* operation call interaction, carrying the *3-Dimensional_Locations* output parameter. Finally, component *Sensor_Data_Acquisition_Daemon* interacts with the *HCCASIS_Database* component through the *SQL_Insert_3-Dimensional_Locations* operation call interaction, carrying the *3-Dimensional_Locations_Query* input parameter.

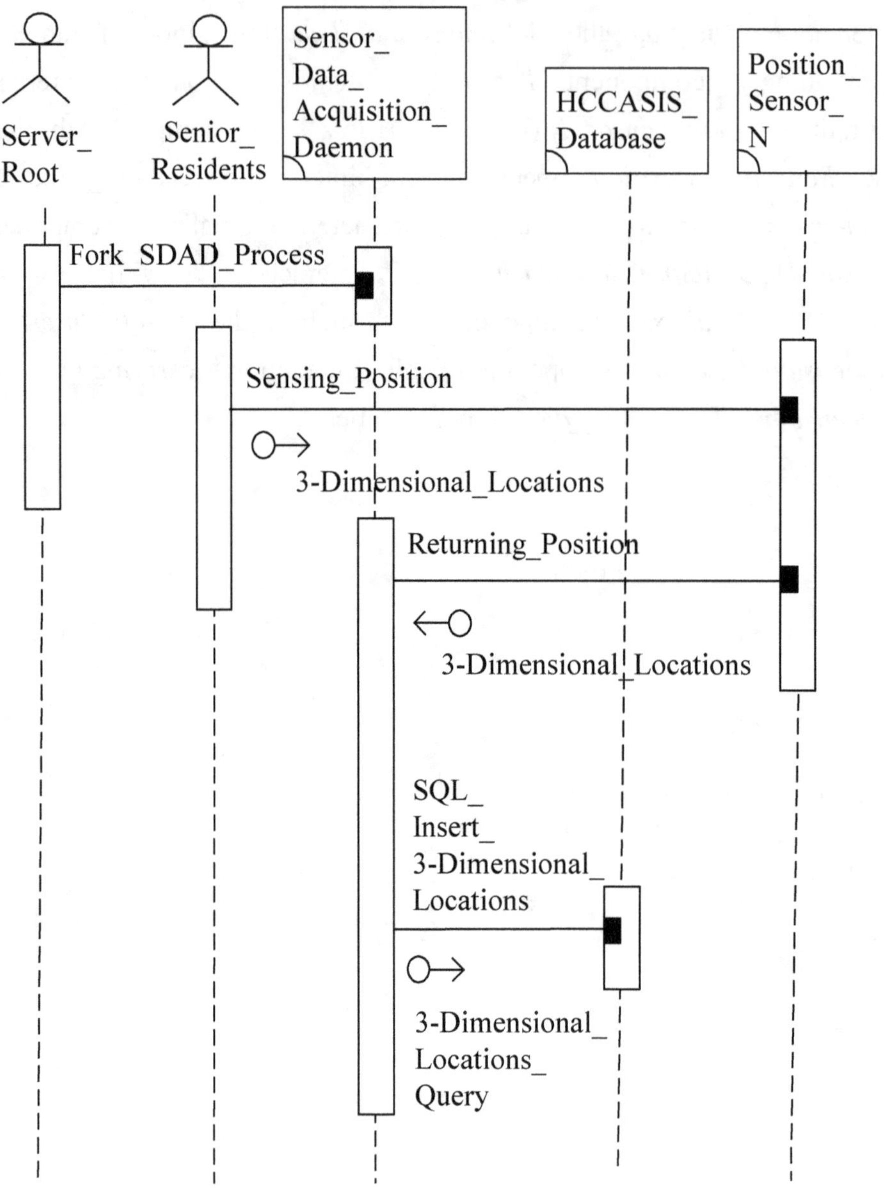

Figure 8-15 IFD of the *Sensing_Residents_Position* Behavior

Figure 8-16 shows an IFD of the *Alerts_Notifying* behavior. First, actor *One_Minute_Interval* interacts with the *Alerts_Notifying_UI* component through the *Showing_All_Alerts* operation call interaction, carrying the *Current_Time* input parameter. Next, component *Alerts_Notifying_UI* interacts with the *HCCASIS_Database* component through the *SQL_Select_3-Dimensional_Locations_for_Alerts_Analysis* operation call interaction, carrying the *Current_Time* input parameter and the *3-Dimensional_Locations_for_Alerts_Analysis_Query* output parameter. Finally, actor *Homecare_Provider* interacts with the *Alerts_Notifying_UI* component through the *Displaying_Alerts* operation call interaction, carrying the *Alerts_Display* output parameter.

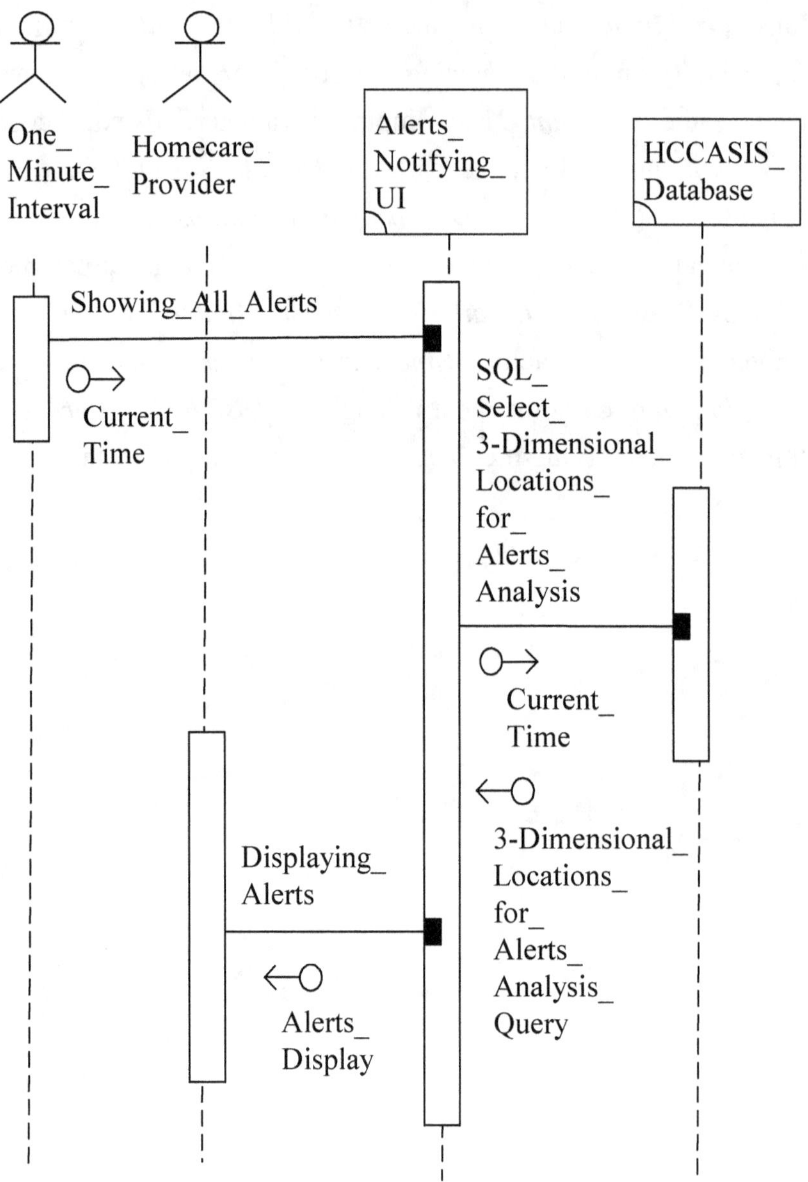

Figure 8-16 IFD of the *Alerts_Notifying* Behavior

Figure 8-17 shows an IFD of the *Recording_Emergency_Responses* behavior. First, actor *Homecare_Provider* interacts with the *Emergency_Responses_UI* component through the *Input_Emergency_Responses* operation call interaction, carrying the *Emergency_Responses_Form* input parameters. Finally, component *Emergency_Responses_UI* interacts with the *HCCASIS_Database* component through the *SQL_Insert_Emergency_Responses* operation call interaction, carrying the *Emergency_Responses_Query* input parameter.

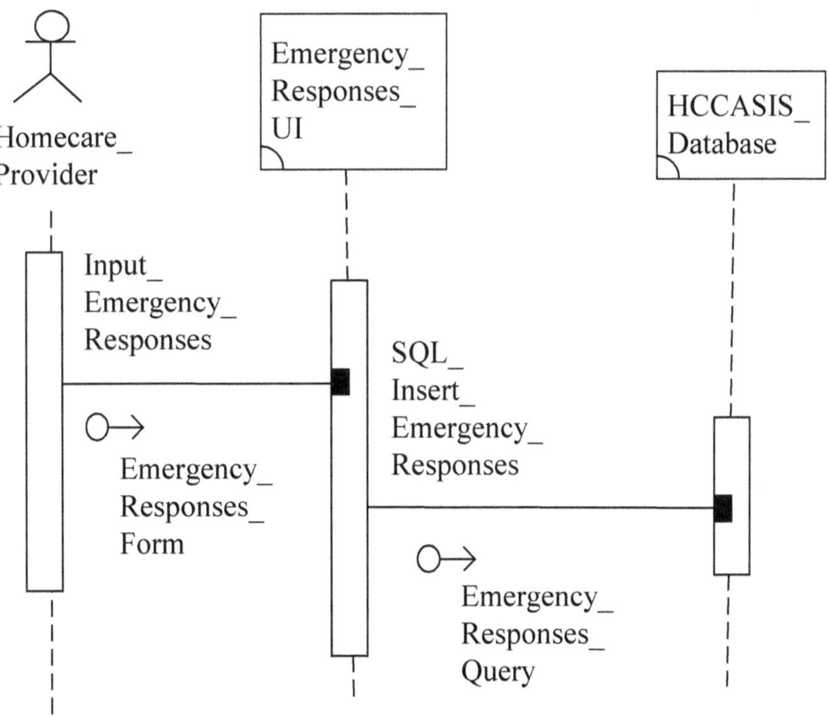

Figure 8-17 IFD of the *Recording_Emergency_Responses* Behavior

Figure 8-18 shows an IFD of the *Printing_Monthly_Statistics* behavior. First, actor *Homecare_Provider* interacts with the *Monthly_Statistics_UI* component through the *PrintButton_Click* operation call interaction, carrying the *Year*, *Month* input parameters. Next, component *Monthly_Statistics_UI* interacts with the *HCCASIS_Database* component through the *SQL_Select_Monthly_Statistics_* operation call interaction, carrying the *Year*, *Month* input parameters and the *Monthly_Statistics_Query* output parameter. Finally, actor *Homecare_Provider* interacts with the *Monthly_Statistics_UI* component through the *PrintButton_Click* operation return interaction, carrying the *Monthly_Statistics_Report* output parameter.

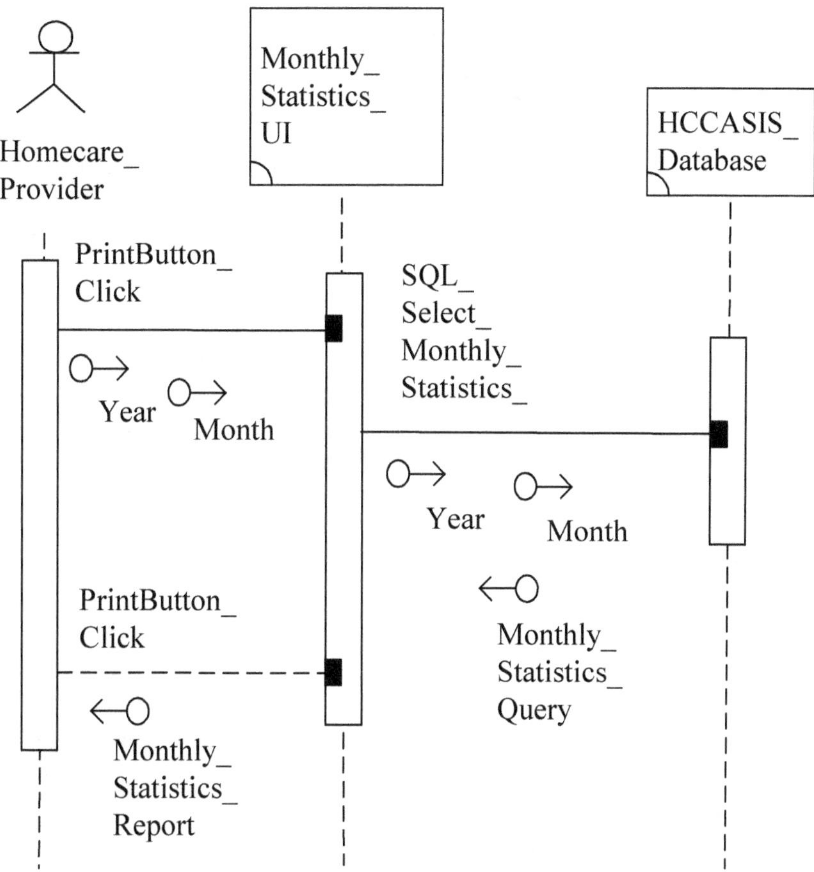

Figure 8-18 IFD of the *Printing_Monthly_Statistics* Behavior

APPENDIX A: SYSTEMS DESIGN 2.0

(1) Framework Diagram

Application Layer

Presentation_Layer

Logic_Layer

Data Layer

Technology Layer

: Component

(2) Component Operation Diagram

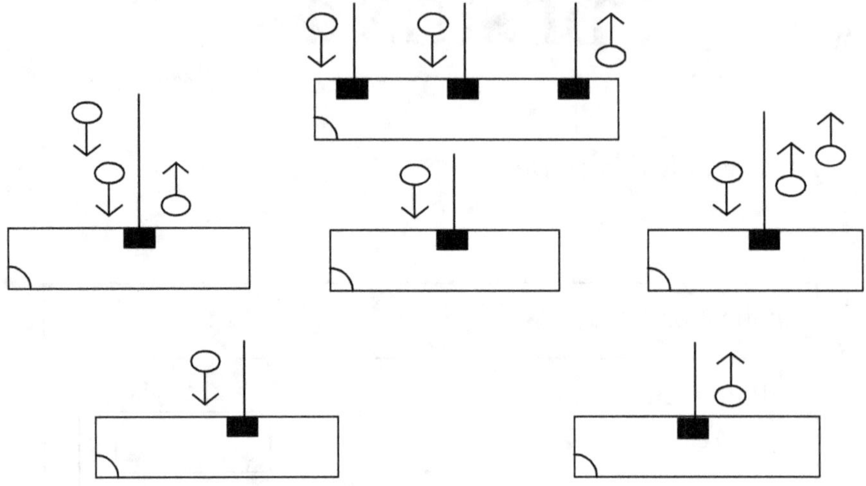

: Operation

: Input Data

: Output Data

: Component

(3) Interaction Flow Diagram

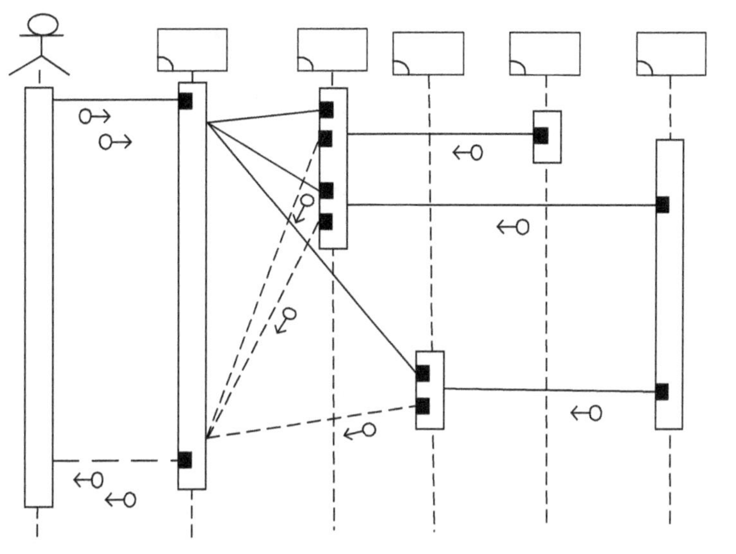

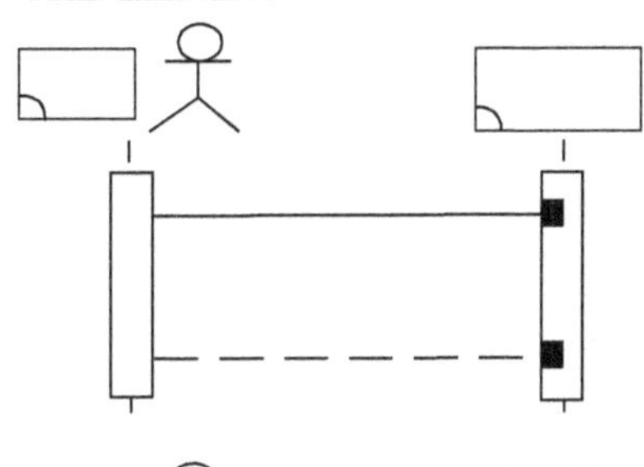

: Operation Call Interaction

: Operation Return Interaction

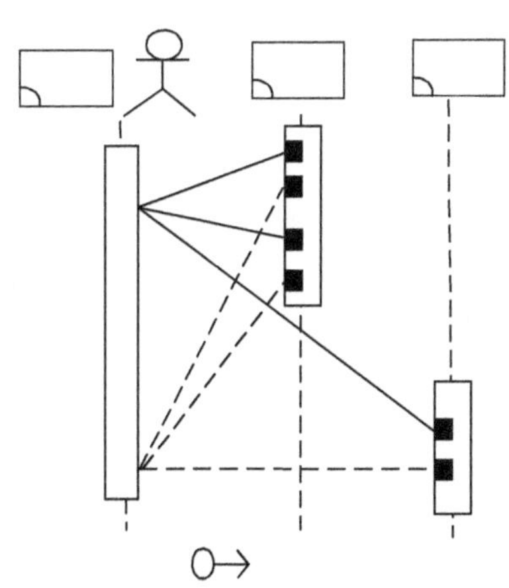

: Conditional
Operation Call Interaction

: Conditional
Operation Return Interaction

: Input Data

: Output Data

APPENDIX B: SBC PROCESS ALGEBRA

(1) Operation-Based Single-Queue SBC Process Algebra

(1) <System> ::= "**fix**(" <Process_Variable> "="<IFD> " • " <Process_Variable>
{"+" <IFD> " • " <Process_Variable>} ")"

(2) <IFD> ::= <Type_1_Interaction> {"• " <Type_1_Or_2_Interaction>}

(3) <Type_1_Or_2_Interaction> ::= <Type_1_Interaction>

| <Type_2_Interaction>

(2) Operation-Based Multi-Queue SBC Process Algebra

(1) <System> ::= <FixIFD> {"|| " <FixIFD>}

(2) <FixIFD> ::= "**fix**(" <Process_Variable>"="<IFD>
"•" <Process_Variable> ")"

(3) <IFD> ::= <Type_1_Interaction> {"• " Type_1_Or_2_Interaction>}

(4) <Type_1_Or_2_Interaction> ::= <Type_1_Interaction>

| <Type_2_Interaction>

(3) Operation-Based Infinite-Queue SBC Process Algebra

(1) <System> ::= "! ("<IFD> " • " *STOP* ")" {"|| ! (" <IFD> " • " *STOP* ")"}

(2) <IFD> ::= <Type_1_Interaction> {"• " <Type_1_Or_2_Interaction>}

(3) <Type_1_Or_2_Interaction> ::= <Type_1_Interaction>

| <Type_2_Interaction>

BIBLIOGRAPHY

[Ashw90] Ashworth, C., *SSADM : A Practical Approach*, 1st Edition, McGraw-Hill Book Company (UK) Ltd., 1990.

[Beam90] Beam, W. R., *Systems Engineering: Architecture and Design*, McGraw-Hill, 1990.

[Bash86] Bashe, C., *IBM's Early Computers*, The MIT Press, 1986.

[Blan08] Blanchard, B. S., *System Engineering Management*, 4th Edition, Wiley, 2008.

[Booc07] Booch, G., *Object-oriented Analysis and Design with Applications*, 3rd Edition, Addison-Wesley, 2007.

[Buha00] Buhalis, D., "Marketing the Competitive Destination of the Future," *Tourism Management*, 2000, pp.97-116.

[Came89] Cameron, John R., *The Jackson Approach to Software Development*, IEEE Computer Society Press, 1989.

[Card11] Cardoso, J. et al., *Reconfigurable Computing: From FPGAs to Hardware/Software Codesign*, Springer, 2011.

[Chao14a] Chao, W. S., *Systems Thingking 2.0: Architectural Thinking Using the SBC Architecture Description Language*, CreateSpace Independent Publishing Platform, 2014.

[Chao14b] Chao, W. S., *General Systems Theory 2.0: General Architectural Theory Using the SBC Architecture*, CreateSpace Independent Publishing Platform, 2014.

[Chao14c] Chao, W. S., *Systems Modeling and Architecting: Structure-*

Behavior Coalescence for Systems Architecture, CreateSpace Independent Publishing Platform, 2014.

[Chao15a] Chao, W. S., *Theoretical Foundations of Structure-Behavior Coalescence*, CreateSpace Independent Publishing Platform, 2015.

[Chao15b] Chao, W. S., *Variants of Interaction Flow Diagrams*, CreateSpace Independent Publishing Platform, 2015.

[Chao15c] Chao, W. S., *A Process Algebra For Systems Architecture: The Structure-Behavior Coalescence Approach*, CreateSpace Independent Publishing Platform, 2015.

[Chao15d] Chao, W. S., *An Observation Congruence Model For Systems Architecture: The Structure-Behavior Coalescence Approach*, CreateSpace Independent Publishing Platform, 2015.

[Chao15e] Chao, W. S., *Variants of SBC Process Algebra: The Structure-Behavior Coalescence Approach*, CreateSpace Independent Publishing Platform, 2015.

[Chao17a] Chao, W. S., *Channel-Based Single-Queue SBC Process Algebra For Systems Definition: General Architectural Theory at Work*, CreateSpace Independent Publishing Platform, 2017.

[Chao17b] Chao, W. S., *Channel-Based Multi-Queue SBC Process Algebra For Systems Definition: General Architectural Theory at Work*, CreateSpace Independent Publishing Platform, 2017.

[Chao17c] Chao, W. S., *Channel-Based Infinite-Queue SBC Process Algebra For Systems Definition: General Architectural Theory at Work*, CreateSpace Independent Publishing Platform, 2017.

[Chao17d] Chao, W. S., *Operation-Based Single-Queue SBC Process Algebra For Systems Definition: General Architectural Theory at Work*, CreateSpace Independent Publishing Platform, 2017.

[Chao17e] Chao, W. S., *Operation-Based Multi-Queue SBC Process Algebra For Systems Definition: Unification of Systems Structure and Systems Behavior*, CreateSpace Independent Publishing Platform, 2017.

[Chao17f] Chao, W. S., *Operation-Based Infinite-Queue SBC Process Algebra For Systems Definition: Unification of Systems Structure and Systems Behavior*, CreateSpace Independent Publishing Platform, 2017.

[Chen76] Chen, P. et al., "The Entity-Relationship Model - Toward a Unified View of Data", *ACM Transactions on Database Systems* 1 (1), pp. 9–36, 1976.

[Date03] Date, C. J., *An Introduction to Database Systems*, 8th Edition, Addison Wesley, 2003.

[DeMa79] DeMarco, T., *Structured Analysis and System Specification*, Prentice Hall, 1979.

[Denn08] Dennis, A. et al., *Systems Analysis and Design*, 4th Edition, Wiley, 2008.

[Dori95] Dori, D., "Object-Process Analysis: Maintaining the Balance between System Structure and Behavior," *Journal of Logic and*

Computation 5(2), pp.227-249, 1995.

[Dori02] Dori, D., *Object-Process Methodology: A Holistic Systems Paradigm*, Springer Verlag, New York, 2002.

[Dori16] Dori, D., *Model-Based Systems Engineering with OPM and SysML*, Springer Verlag, New York, 2016.

[Elma10] Elmasri, R., *Fundamentals of Database Systems*, 6th Edition, Addison Wesley, 2010.

[Ghar11] Gharajedaghi, J., *Systems Thinking: Managing Chaos and Complexity: A Platform for Designing Business Architecture*, Morgan Kaufmann, 2011.

[Grad13] Grady, J. O., *System Requirements Analysis*, 2nd Edition, Elsevier, 2013.

[Hatl00] Hatley, D. J. 2t al., *Process for System Architecture and Requirements Engineering*, 1st Edition, 2000.

[Hoar85] Hoare, C. A. R., *Communicating Sequential Processes*, Prentice-Hall, 1985.

[Hoff10] Hoffer, J. A., et al., *Modern Systems Analysis and Design*, Prentice Hall, 6th Edition, 2010.

[Kend10] Kendall, K. et al., *Systems Analysis and Design*, 8th Edition, Prentice Hall, 2010.

[Koss11] Kossiakoff, A. et al., Systems Engineering Principles and Practice, 2nd Edition, Wiley-Interscience, 2011.

[Marc88] Marca, D. A. et al., *SADT: Structured Analysis and Design Technique,* McGraw-Hill, 1988.

[Miln89] Milner, R., *Communication and Concurrency*, Prentice-Hall, 1989.

[Miln99] Milner, R., *Communicating and Mobile Systems: the π-Calculus*, 1st Edition, Cambridge University Press, 1999.

[Pask09] Paskaleva, K., "Enabling the Smart City: The Progress of E-City Governance in Europe," *International Journal of Innovation and Regional Development*, 2009, pp.405-422.

[Pele00] Peleg, M. et al., "The Model Multiplicity Problem: Experimenting with Real-Time Specification Methods". *IEEE Tran. on Software Engineering*. 26 (8), pp. 742–759, 2000.

[Prat00] Pratt, T. W. et al., *Programming Languages: Design and Implementation*, 4th Edition, Prentice Hall 2000.

[Pres09] Pressman, R. S., *Software Engineering: A Practitioner's Approach*, 7th Edition, McGraw-Hill, 2009.

[Reis92] Reisig, W., A Primer in Petri Net Design, Springer-Verlag, 1992.

[Scho10] Scholl, C., *Functional Decomposition with Applications to FPGA Synthesis*, Springer, 2010.

[Seth96] Sethi, R., *Programming Languages: Concepts and Constructs*, 2nd Edition, Addison-Wesley, 1996.

[Shel11] Shelly, G. B., et al., *Systems Analysis and Design*, 9th Edition, Course Technology, 2011.

[Shyu02] Shyu, L. et al., "Predictors of Nursing Home Placement and Home Nursing Services Utilization by Elderly Patients after Hospital Discharge in Taiwan," *Journal of Advanced Nursing*, 2002, pp.398-406.

[Sode03] Soderborg, N.R. et al., "OPM-based Definitions and Operational Templates," *Communications of the ACM* 46(10), pp. 67-72, 2003.

[Somm06] Sommerville, I., *Software Engineering*, 8th Edition, Addison-Wesley, 2006.

[Wald15] Walden, D. D. et al., *INCOSE Systems Engineering Handbook: A Guide for System Life Cycle Processes and Activities*, 4th Edition, Wiley, 2015.

INDEX

N

O

P

S

www.ingramcontent.com/pod-product-compliance
Lightning Source LLC
Chambersburg PA
CBHW081454170526
45166CB00008B/2425

* 9 7 8 1 5 2 2 7 8 6 2 0 7 *